Essential

A2 Chemistry

for OCR

Ted Lister
Janet Renshaw

Published in 2004 by:
Nelson Thornes Ltd
Delta Place
27 Bath Road
CHELTENHAM
GL53 7TH
United Kingdom

04 05 06 07 08 / 10 9 8 7 6 5 4 3 2 1

A catalogue record for this book is available from the British Library

ISBN 0 7487 8506 X

Illustrations by IFA Design and Oxford Designers & Illlustrators
Page make-up by Tech Set

Printed and bound in Croatia by Zrinski

Contents

For Joseph, Thomas, Georgia, Ella and Louis – chemists of the future?

Introduction

This book has been written to meet the Oxford, Cambridge and RSA (OCR) specification for Advanced Level (A2) Chemistry. It matches the specification and has therefore been endorsed by OCR. The subject content is divided into three modules (**Chains, rings and spectroscopy, Trends and patterns** and **Unifying concepts**) that correspond to the three modules of the OCR specification) plus an appendix, **Mathematics in Chemistry**. Each module is subdivided into chapters that broadly correspond to the sub-divisions of these modules in the OCR specification.

Its features include:
- **Units** each covering a single topic and dividing the content into manageable portions.
- **Numbered sections** to further divide the content into bite-sized pieces and to allow easy cross referencing.
- **Extensive use of bullet points** to produce lists of information that make learning and revision easier.
- **Full colour diagrams** to aid understanding and improve clarity.
- **Colour photographs** for extra impact and realism.
- **Accessible language** to make understanding easier.
- **Bold type** to emphasise key words and ideas.

- **Purple type** to refer to explanations and definitions in the glossary.
- **Comprehensive glossary** of explanations and definitions.
- **Extensive cross referencing** to link different topics and allow an understanding of chemistry as a whole.
- **Quick questions** after each topic to test your understanding.
- **OCR exam questions** after each module to test the skills expected at A2 level including application of knowledge, understanding, analysis, synthesis and evaluation.

There is also a **CD ROM** of teacher support material and resources including extra support for students to be used alongside the book, answers to the OCR exam questions (with examiners' commentary), help with practical work, links to useful websites and much more.

We hope that you will enjoy using this book and that it will lead to success in your AS Chemistry examinations.

Ted Lister and Janet Renshaw

Acknowledgements

The authors would like to thank John Payne for helpful comments and discussions on the manuscript and for advice on syllabus matching.

The authors would also like to thank the team at Nelson Thornes for their help in producing this book – Beth Hutchins, commissioning editor; John Hepburn and Marilyn Grant, editors; Holly Myers, publishing assistant and John Bailey and Stuart Sweatmore, picture research.

The authors and publishers are grateful to Oxford, Cambridge and RSA Examinations for kind permission to reproduce examination questions.

Photograph acknowledgements

Alamy: 56 David Hoffman Photolibrary, 115 Agence Images; Allsport: 46B Agence Vandystadt/Jean-Marc Labout; Avesta: 88h; BASF: 34; BMW Rover Group: 88b; Copper Pipe & Plumbing Association: 86TR; Corel (NT): 42T, 128; Corel 243 (NT): 88c; Corel 30 (NT): d; Corel 61 (NT): e; Ingram IL V1 CD3 (NT): 88f; Jeyes: 18BL; Johnson & Johnson: 18BR; Martyn Chillmaid: 11, 23, 26, 30, 40, 42B, 76, 82, 88a, 91, 124L; Oxford Scientific Films: 19 Scott Camazine, 122 Densey Cline; Photodisc 28 (NT): 86TL; Proctor & Gamble 129L, R; Reckitt Benckiser: 18T; RV Cool: 86MR; Science Photolibrary: 15 Martyn Chillmaid, 46T, 77R Charles D Winters, 57 Mauro Fermariello, 77L, 84, 86B, 87, 88BR, 109, 118, 124R Andrew Lambert Photography, 116; Tim Clayton: 86ML.

Every effort has been made to trace all the copyright holders but if any have been overlooked the publisher will be pleased to make the necessary arrangements at the first opportunity.

Chains, rings and spectroscopy

Arenes

Introduction to arenes

1.1.1 Introduction

Arenes are hydrocarbons, based on benzene, C_6H_6, which is the simplest. Benzene has a hexagonal (six-sided) ring structure that is very stable because it has a special type of bonding. Arenes were first isolated from sweet-smelling oils, such as balsam, and this gave them the name **aromatic** compounds. Arenes are still called aromatic compounds but this now refers to their structures rather than their aromas. Benzene and other arenes have characteristic properties.

Benzene is given the special symbol
benzene

Arenes can have two or more rings fused together as in the examples below.

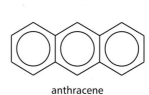  

naphthalene   anthracene   phenanthrene

Arenes can have other functional groups (substituents) replacing one or more of the hydrogen atoms in their structures.

1.1.2 Bonding and structure of benzene

The bonding and structure of benzene C_6H_6 was a puzzle for a long time to organic chemists, because:

- benzene was more stable than expected;
- benzene reacted by substitution rather than addition;
- in benzene all the carbon atoms were equivalent, which implied that all the carbon–carbon bonds are the same.

The box *Early ideas on the structure of benzene*, on page 8, describes how the puzzle began to be solved.

Benzene consists of a flat, regular hexagon of carbon atoms, each of which is bonded to a single hydrogen atom. The geometry of benzene is shown in Figure 1.1.

The C—C bond lengths in benzene are intermediate between those expected for a carbon–carbon single bond and a carbon–carbon double bond, see Table 1.1. So, each bond is intermediate between a single and a double bond.

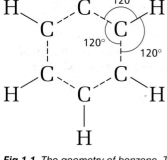

Fig 1.1 *The geometry of benzene. The dashed lines show the shape and do not represent single bonds.*

Table 1.1 *Carbon–carbon bond lengths*

Bond	Length/nm
C—C	0.154
C═══C (in benzene)	0.140
C=C	0.134
C≡C	0.120

The symbol 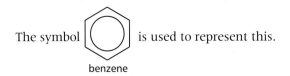 is used to represent this.

The carbon atoms are bonded by σ-bonds and there is also a molecular orbital, shaped like a doughnut, above and below the ring, extending over all six carbon atoms. This has been made by the overlap of p-orbitals – one from each carbon atom, see Figure 1.2. The electrons in this orbital are **delocalised** and form a π-bond. Delocalisation means that electrons are spread over more than two atoms – in this case the six carbon atoms that form the ring, see Figure 1.2. Overall, each carbon–carbon bond has a total of three electrons, making it intermediate between a single and a double bond. The delocalised system is very important in the chemistry of benzene and its derivatives. It makes benzene unusually stable. This is sometimes called aromatic stability, Figure 1.3.

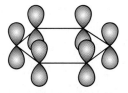

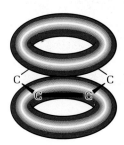

Fig 1.2 *Overlap of p-orbitals to form the π-bond in benzene*

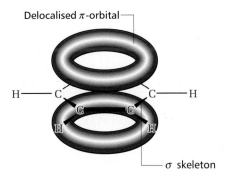

Fig 1.3 *The delocalised π-orbital in benzene*

A technique called X-ray diffraction shows a contour map of the electron density in an individual benzene molecule. This shows that the benzene molecule is a perfect hexagon and each carbon–carbon bond length is 0.140 nm.

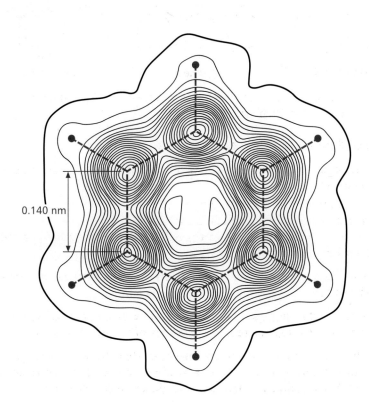

0.140 nm

EARLY IDEAS ON THE STRUCTURE OF BENZENE

The formula of benzene, C_6H_6, suggests that its structure should have a lot of double bonds, It could, for example, look like:

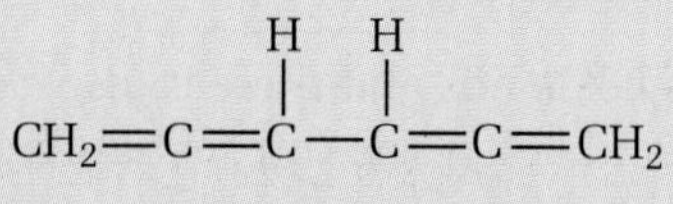

$$CH_2=C=C-C=C=CH_2$$

But, this **unsaturated** compound, an alkene, would easily undergo addition reactions across the double bonds and benzene does not do this – it almost always reacts by substitution rather than addition.

In 1865 Friedrich August Kekulé proposed a ring structure. The idea came to him in a dream about snakes swallowing their own tails!

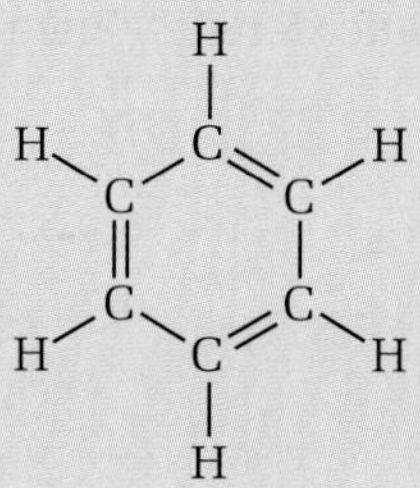

or 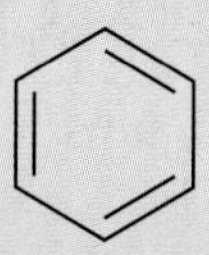   in skeletal notation

Even this structure did not account for all the properties of benzene. For example:

In this hexagonal structure the carbon–carbon bonds would not all be of equal length because double bonds are shorter than single bonds.

It would still show addition reactions because it is an alkene.

Kekulé then proposed that benzene consisted of two structures in rapid equilibrium. He suggested that it was this 'resonating structure' that gave the ring its stability. He had almost reached the right answer.

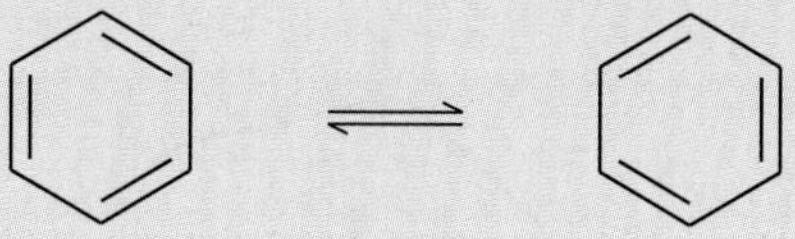

QUICK QUESTIONS

1 What is the empirical formula of benzene?

2 What is **a** the molecular and **b** the empirical formula of naphthalene? You might need to draw all the hydrogen atoms onto the skeletal formula and think carefully about what the ring in the structure means.

3 **a** How many molecules of hydrogen, H_2, could add on to a benzene molecule to give a fully saturated product cyclohexane?

   b How many molecules of hydrogen, H_2, could add on to the structure shown at the top of the box *Early ideas on the structure of benzene*?

   c Explain the difference in your answers to **a** and **b**.

4 Explain what is meant by 'delocalisation' of electrons in the benzene ring.

Arenes – physical properties, naming and reactivity

1.2.1 Physical properties of arenes

Benzene is a colourless liquid at room temperature. It boils at 353 K and freezes at 279 K. Its boiling temperature is comparable with that of hexane (342 K) but its freezing temperature is much higher than hexane's (178 K). This is because benzene's flat, hexagonal molecules pack together very well in the solid state, Figure 1.4. They are therefore harder to separate and this must happen for the solid to melt.

Like other hydrocarbons that are non-polar, arenes do not mix with water, but they do mix with other hydrocarbons and other non-polar solvents.

1.2.2 Naming aromatic compounds

Substituted arenes are generally named as derivatives of benzene. So that benzene forms the root of the name.

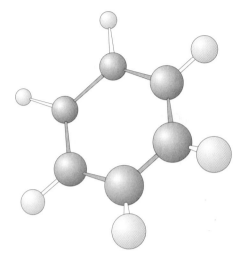

$C_6H_5CH_3$, is called methylbenzene;

C_6H_5Cl, is called chlorobenzene and so on.

However,

C_6H_5OH, is still called phenol (its original name) and *not* benzeneol.

When there is more than one substituent, locants are used to show where they are attached. So:

$C_2H_4NO_2OH$, is called 2-nitrophenol.

Fig 1.4 *Benzene molecules, top, can pack together better than hexane molecules, bottom, so benzene has a higher freezing temperature than hexane*

However, benzene rings with two —OH groups are named systematically, using locants to place the —OH groups:

$C_6H_4(OH)_2$, is benzene-1,2-diol.

It is more important to be able to work from the name to the correct structure than to correctly name a compound whose formula is given.

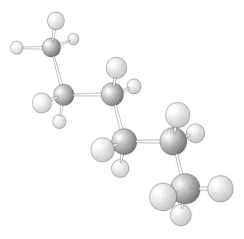

HINT

Benzene itself is carcinogenic (may cause cancer) so in your practical work in school or college you are likely to use related compounds that are safer, such as methylbenzene.

9

HINT

Non-systematic names are still used for many derivatives of benzene. In particular benzenecarboxylic acid is nearly always called benzoic acid and benzenecarbaldehyde is called benzaldehyde.

HINT

An electrophile has a positive charge – either as a positive ion or the positive end of a dipole.

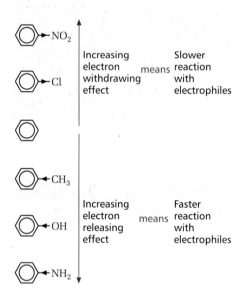

Fig 1.5 *Electron-withdrawing and electron-releasing substituents*

Examples

You can test yourself by covering the names or the structures:

Ethylbenzene $\qquad$ $C_6H_5C_2H_5$

Nitrobenzene $\qquad$ $C_6H_5NO_2$

1,2-dimethylbenzene $\qquad$ $C_6H_4(CH_3)_2$

1.2.3 The reactivity of aromatic compounds

Two factors are important to the reactivity of aromatic compounds.

- The ring is an area of high electron density, because of the delocalised bond, see Section 1.1.2, and is therefore is attacked by **electrophiles**.
- The aromatic ring is very stable. It needs energy to be put in to break the ring before the system is destroyed. We call this the 'aromatic stabilisation energy'. It means that the ring almost always remains intact in the reactions of arenes.

The above two points mean that most of the reactions of aromatic systems are **electrophilic substitution** reactions.

The effect of substituents

If a benzene ring already has one or more substituents, they may affect further electrophilic substitution reactions. They may withdraw electrons from the ring making it less reactive or release electrons onto the ring making it more reactive, Figure 1.5.

QUICK QUESTIONS

1 The electronegativity of carbon is 2.5 and that of hydrogen is 2.1. Explain why this makes benzene almost non-polar.

2 What intermolecular forces act between non-polar molecules?

3 Name **a** [structure], **b** [structure].

4 Draw the structure of **a** 1,4-dimethylbenzene, **b** 2-chlorophenol.

5 Which of the following is an electrophile?
H^+, $:NH_3$, Cl_2, $Cl\cdot$, Cl^-.

Reactions of arenes

1.3.1 Combustion

Arenes burn in air with flames that are noticeably smoky. This is because they have a high carbon:hydrogen ratio. There is usually unburned carbon remaining when they burn in air and this produces soot. A smoky flame suggests an aromatic compound.

1.3.2 Electrophilic substitution reactions

Although benzene is unsaturated it does not react like an alkene. The reaction with bromine, see below, Section 1.3.3, highlights the difference.

The reaction is an electrophilic substitution that leaves the aromatic system unchanged, rather than addition, which would require the input of the aromatic stabilisation energy to destroy the aromatic system.

The mechanism of electrophilic substitutions

The delocalised system of the aromatic ring has a high electron density that attracts electrophiles. The electrons are attracted towards the electrophile, El^+.

A bond forms between one of the carbon atoms and the electrophile. But, to do this, the carbon must use electrons from the delocalised system. This destroys the aromatic system. To get back the stability of the aromatic system, the carbon loses a H^+ ion. The sum of these reactions is the substitution of H^+ by El^+.

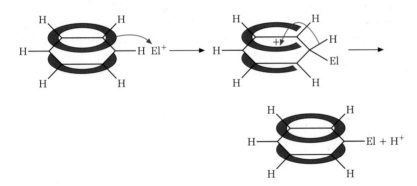

The same overall process occurs in halogenation, nitration and Friedel–Crafts reactions.

1.3.3 Halogenation

Halogenation is the substitution of a halogen atom for one of the hydrogen atoms on the aromatic ring. Halogen molecules can act as electrophiles because they have instantaneous dipoles $X^{\delta+}-X^{\delta-}$. However, they are not good enough electrophiles to react with benzene except in the presence of a catalyst. Suitable catalysts are aluminium halides or iron filings (which react with halogens to form iron(III) halides). The catalysts work as halogen 'carriers' as shown below, using bromine and iron filings as an example:

$$\overset{\delta+}{Br}-\overset{\delta-}{Br}\text{:} + \overset{\delta+}{Fe}\overset{\nearrow Br^{\delta-}}{\underset{\searrow Br^{\delta-}}{-Br^{\delta-}}} \rightleftharpoons Br^+ + FeBr_4^-$$

Arenes burn with a smoky flame

11

A small concentration of the electrophile Br^+ is formed which then attacks the benzene ring. The bond is first formed between Br^+ and one of the carbon atoms:

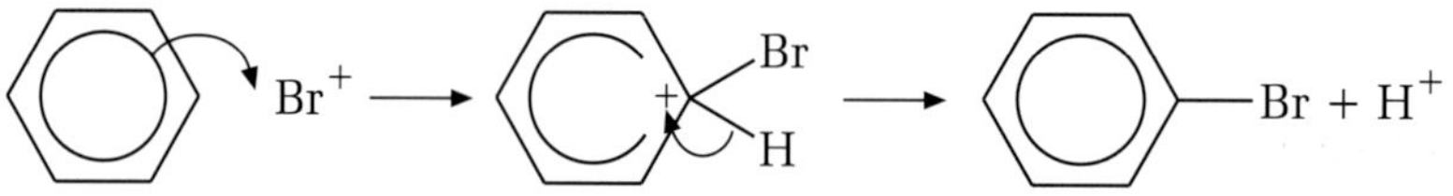

An H^+ ion is then lost.

The overall reaction is:

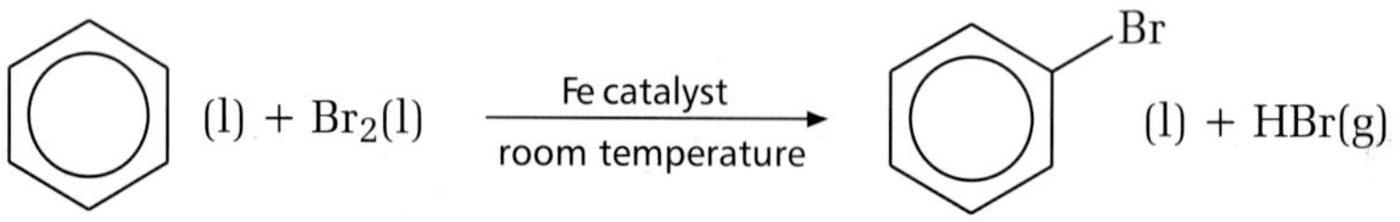

The hydrogen bromide forms white fumes on contact with moist air. Its presence confirms that hydrogen has been removed from the benzene and so the reaction must be *substitution* rather than *addition*.

To substitute chlorine, we use chlorine and aluminium chloride, $AlCl_3$, which works as a halogen carrier catalyst in the same way as iron(III) bromide, $FeBr_3$. It generates the electrophile Cl^+ (and $AlCl_4^-$).

$$C_6H_6(l) + Cl_2(g) \xrightarrow{AlCl_3 \text{ catalyst}} C_6H_5Cl(l) + HCl(g)$$

Halogen atoms are more **electronegative** than carbon atoms, so when bonded to a benzene ring they attract electrons away from it. So halogenoarenes are less susceptible to further attack by electrophiles. This means that very little of the disubstituted products, such as 1,2-dihalobenzene and 1,4-dihalobenzene, are formed.

Notice the difference between the reaction of benzene with bromine (which is a substitution) and an alkene, such as cyclohexene, with bromine (which is an addition). Almost all the reaction of benzene are substitutions – addition reactions would destroy the delocalised aromatic system of the benzene ring.

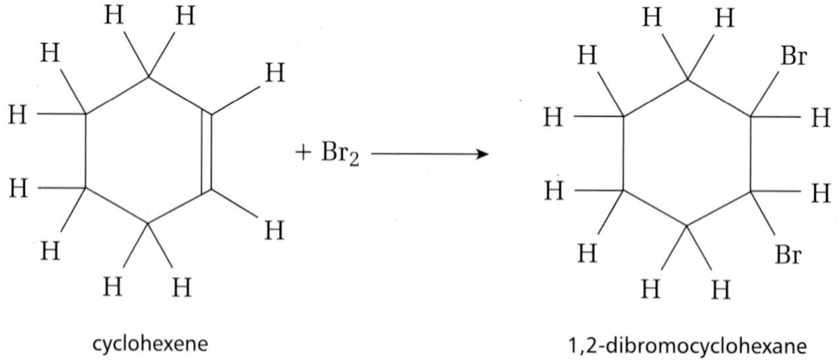

cyclohexene   1,2-dibromocyclohexane

This is an addition reaction. It does not need a catalyst and takes place with a dilute solution of bromine. The bromine solution is decolourised.

Benzene reacts only with liquid bromine (much more concentrated) and only with the help of a catalyst.

In the pharmaceutical industry chlorobenzene is converted into phenylamine, $C_6H_5NH_2$, which is used in the manufacture of a number of drugs.

HINT

Neither bromination nor chlorination need ultraviolet (UV) light, although halogens will *add on* to arenes in UV light.

QUICK QUESTIONS

1 Write the equation for the complete combustion of benzene in oxygen.

2 Suggest which of the following would be least susceptible to electrophilic attack – chlorobenzene, bromobenzene or iodobenzene.

3 Explain your answer to question 2.

4 Why are most of the reactions of benzene, substitutions rather than additions?

More reactions of arenes

1.4.1 Nitration

Nitration is the substitution of a —NO_2 group for one of the hydrogen atoms on an arene ring. It has the same mechanism as the one we saw in Section 1.3.2. The electrophile NO_2^+ is generated in the reaction mixture of concentrated nitric and sulphuric acids:

$$HNO_3 + H_2SO_4 \rightarrow NO_2^+ + H_2O + 2HSO_4^-$$

NO_2^+ acts as an electrophile and the following mechanism occurs:

The overall product of the reaction of the reaction of the NO_2^+ with benzene is nitrobenzene:

nitrobenzene

The balanced equation is;

Notice that sulphuric acid does not appear; in effect it acts as a catalyst.

A little dinitrobenzene may also be formed by the further attack of NO_2^+ on nitrobenzene. The extra NO_2 group goes to the 3-position, so the di-substituted isomer produced is 1,3-dinitrobenzene.

1,3-dinitrobenzene

To produce nitrobenzene, benzene is warmed with a mixture of nitric and sulphuric acids at 330 K, see Figure 1.6. 1,3-dinitrobenzene is solid at room temperature and is separated from the liquid nitrobenzene by cooling the mixture.

> **HINT**
>
> H_2SO_4 is a stronger acid than HNO_3 and donates a proton (H^+) to HNO_3, which then loses a molecule of water to give NO_2^+.

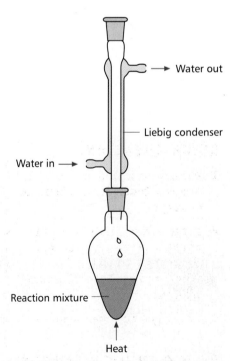

Fig 1.6 *Apparatus for preparing nitrobenzene. The vertical condenser ensures that any benzene vapour condenses and is returned to the reaction flask*

13

1.4.2 Friedel–Crafts reactions

These reactions substitute an organic group, R, for a hydrogen atom on an aromatic ring. They use aluminium chloride as a catalyst and were discovered by Charles Friedel and James Crafts.

Halogenoalkanes provide the alkyl group. They react with $AlCl_3$ to form $AlCl_4^-$ and R^+:

$$RCl + AlCl_3 \rightarrow R^+ + AlCl_4^-$$

R^+ is a good electrophile that attacks the benzene ring to form substitution products in the same way as other electrophiles.

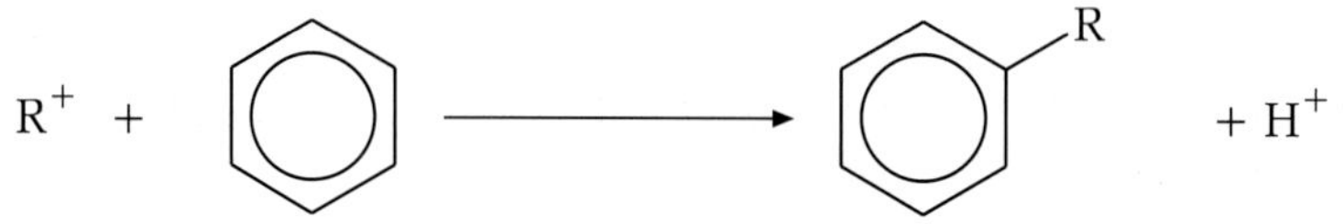

The products are alkyl substituted arenes The overall reactions are:

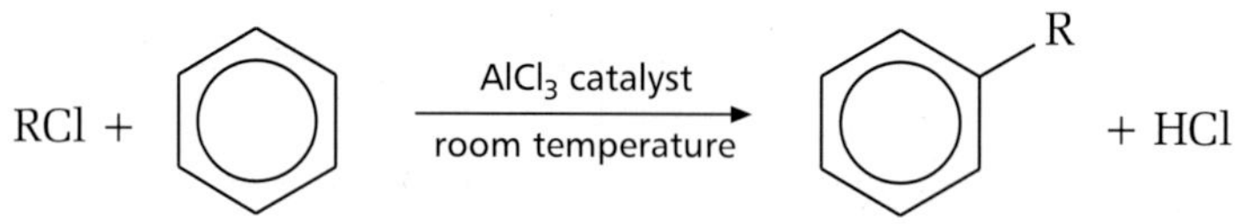

For example, chloromethane reacts with benzene to form methylbenzene:

$$CH_3Cl + AlCl_3 \rightarrow CH_3^+ + AlCl_4^-$$

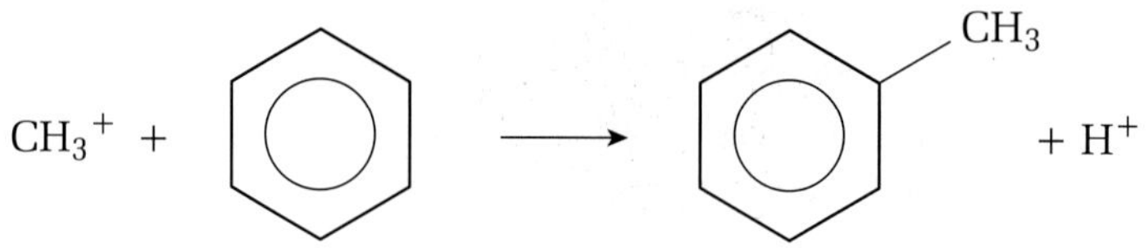

methylbenzene

Notice the similarity between the way the aluminium chloride acts as a catalyst and the action of the iron(III) bromide on the halogenation reaction, in Section 1.3.3. Both act by generating a good electrophile.

In industry, alkylbenzenes are produced as the first stage in the process for making detergents.

HINT

In organic chemistry, R— is used to indicate an unspecified organic group. Here it represents an alkyl group, such as methyl, CH_3—, ethyl, C_2H_5—, etc.

QUICK QUESTIONS

1 Classify (i) nitration, (ii) Friedel–Crafts reactions as:
   a electrophilic substitution,
   b nucleophilic substitution,
   c electrophilic addition,
   d free-radical addition,
   e free radical substitution.

2 Write an equation for the Friedel–Crafts reaction that produces ethylbenzene.

3 What species attacks the benzene ring in the reaction in question 2?

4 Name the two isomers of 1,3-dinitrobenzene.

Phenols

Phenols are aromatic compounds in which one or more of the hydrogen atoms of the ring are replaced by an —OH group. Phenols still have the aromatic π-system.

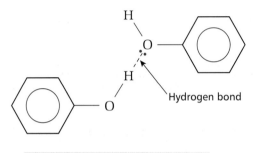

C_6H_5OH, phenol

phenol

Phenol itself is a pale pink crystalline solid. The hydrogen bonds between phenol molecules cause phenol to be a solid.

Hydrogen bond

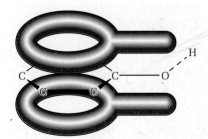

Phenol crystals

1.5.1 Comparing phenols with alcohols

Although phenols and alcohols both contain the —OH functional group they behave very differently. Alcohols tend to react by substitution of the —OH group with another **nucleophile** – the C—OH bond breaks. In phenols the —OH group is more firmly attached to the carbon on the ring. This is because of the influence of the aromatic ring. The bond most likely to break is therefore the O—H bond.

1.5.2 Bonding in phenols

The bonding in the benzene ring interacts with that in the —OH group. Figure 1.7 shows how one of the lone pairs on the oxygen atom in phenol can overlap with the delocalised π-system on the benzene ring to form an extended delocalised π-orbital.

As a result of this delocalisation, the C—O bond in phenol has some double bond character and is stronger than the C—O single bond in an alcohol, but not as strong as a C=O double bond. We can see this by comparing their bond lengths, see Table 1.2

The extra strength of the bond means the C—O bond in phenols is harder to break than the C—O bond in alcohols.

A second reason that the C—O bond in phenol does not break as easily as the same bond in alcohols, is also the result of the overlapping of π-orbitals.

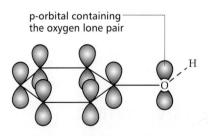

p-orbital containing the oxygen lone pair

Fig 1.7 *Interaction of the oxygen lone pair with the aromatic system in phenol*

Table 1.2 *Comparison of bond lengths. Remember, the shorter the bond, the stronger it is*

Bond	Bond length/nm
Carbon–oxygen double bond	0.122
Carbon–oxygen single bond	0.143
Carbon–oxygen bond in phenol	0.136

HINT

An electronegative atom attracts negative charge.

Carbon atoms bonded to oxygen atoms always have a $\delta+$ charge, because oxygen is more electronegative than carbon. In phenol, this $\delta+$ charge is spread over the whole delocalised system. This makes this carbon less $\delta+$ and therefore less susceptible to attack by nucleophiles than is the C—O bond in alcohols.

For these two reasons loss of the —OH group is less likely in phenols than in alcohols, see Figure 1.8.

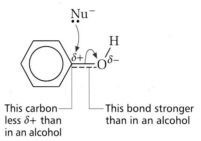

Fig 1.8 *Phenol is less susceptible than alcohols to attack by nucleophiles*

1.5.3 Reactivity

Phenols will react fairly readily by breaking the O—H bond, and losing a H^+ ion. This means that:

- Phenols are somewhat acidic. When phenol loses a proton (H^+) ion we get a

phenoxide ion ⬡—O^-.

The phenoxide ion is relatively stable because of the overlap of the π-orbitals on the oxygen with the π-system of the benzene ring – the negative charge of this ion is delocalised over the whole ion. Alkoxide ions such as ethoxide, $C_2H_5O^-$, which forms when ethanol loses a hydrogen ion, are very unstable because no delocalisation is possible.

1.5.4 The effect of —OH on the benzene ring

As well as reactions of the —OH group, the benzene ring itself can also react. The —OH group activates the ring by releasing electrons onto it. Although oxygen attracts electrons through the C—O σ-bond, this is outweighed by the release of electrons from oxygen's lone pairs onto the benzene ring through the π-overlap. This means that phenol is more reactive towards electrophilic substitution than benzene itself.

QUICK QUESTIONS

1 Phenol is acidic. Write equations for the reaction of phenol with

   a sodium hydroxide,

   b sodium.

2 Which of the reactions in question 1 will not take place with ethanol?

3 Phenol cannot be oxidised to a carbonyl compound or a carboxylic acid. In this respect, what type of alcohol does it most resemble – primary, secondary, tertiary?

4 Aqueous bromine solution reacts with phenol at room temperature to give 2,4,6-tribromophenol:

   a Draw the structural formula of 2,4,6-tribromophenol;

   b What will be the other (inorganic) product of the reaction?

   c Look up the conditions under which benzene reacts with bromine (Topic 1.3) and the product.

   d Give three ways in which this reaction shows that phenol is more reactive towards electrophiles than benzene.

Reactions of phenol

1.6.1 Combustion

Phenol burns in air with the smoky flame characteristic of aromatic compounds.

1.6.2 Reactions of the —OH group in phenol

Acidity
Phenol is a weak acid. When it loses a H^+ ion it forms a phenoxide ion. This is relatively stable owing to the delocalisation of the negative charge, see Section 1.5.3.

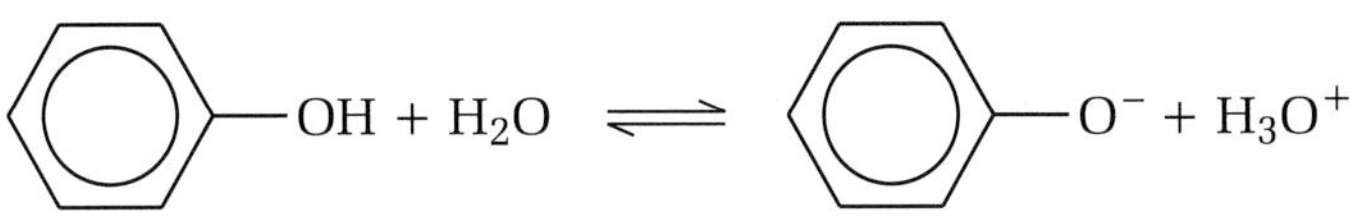

phenoxide ion

It therefore has some of the typical acid reactions:

Reaction with alkalis
Phenol is only slightly soluble in water but dissolves in aqueous sodium hydroxide solution to form the salt sodium phenoxide and water.

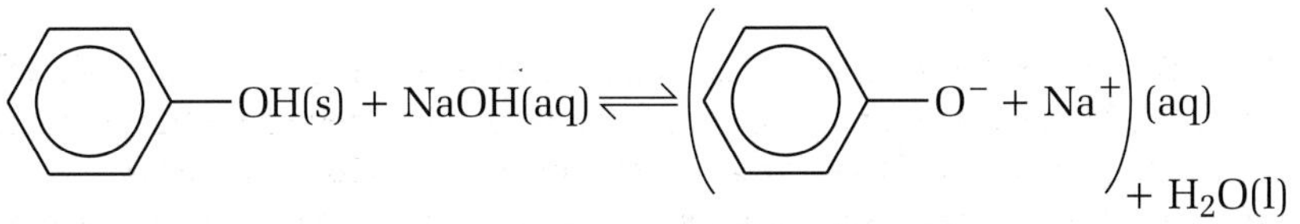

sodium phenoxide

> **HINT**
> This is an example of acid + alkali → salt + water

The equilibrium can be moved to the left by adding concentrated hydrochloric acid, which makes the phenol precipitate out of solution.

Reaction with alkali metals
Alkali metal phenoxides can also be made by direct reaction of molten phenol with the alkali metal. Hydrogen is given off. For example, with sodium, we get sodium phenoxide and hydrogen.

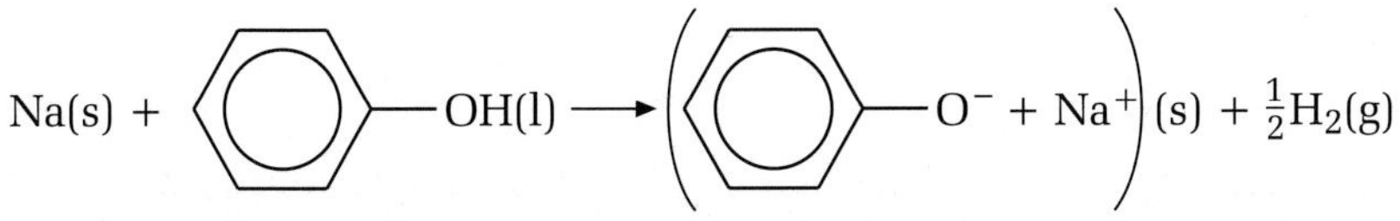

> **HINT**
> This is an example of a typical acid reaction: acid + metal → salt + hydrogen

Phenol is not a strong enough acid to react with sodium carbonate solution to produce carbon dioxide.

1.6.3 Reactions of the benzene ring in phenol

The —OH group releases electrons onto the aromatic ring, so the aromatic ring in phenol is much more reactive towards electrophilic substitution than benzene. This may be shown in various ways:

- The same reaction occurs more rapidly.
- More substitution takes place under the same conditions.
- Less vigorous conditions are needed for the same reaction.

> **DID YOU KNOW?**
> Phenol was once known as carbolic acid. The name is still used in carbolic soaps, which contain small amounts of phenol, derivatives as antiseptics. (Pure phenol is an unpleasantly corrosive substance causing blistering of the skin and it should be treated with respect.)

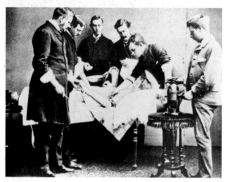

Phenol (carbolic) sprays were used in early operations

Halogenation

The —OH group directs further substituents to the 2-, 4- and 6-positions. For example, phenol will react at room temperature with aqueous bromine or chlorine to give a trisubstituted product. No catalyst is needed. With bromine solution we produce 2,4,6-tribromophenol:

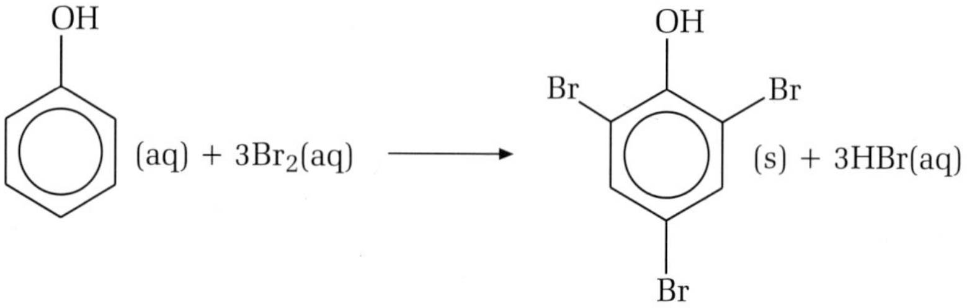

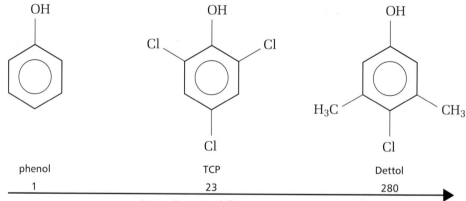

A white precipitate of 2,4,6-tribromophenol is produced and the brown colour of the bromine disappears.

Compare this with benzene. Liquid bromine (rather than aqueous bromine) is needed and a catalyst is required, yet the product is bromobenzene, in which only one of the hydrogen atoms has been substituted by bromine.

1.6.4 Phenols in industry

Antiseptics and disinfectants both kill germs. Antiseptics may be applied to the skin while disinfectants are normally only applied to surfaces. Phenol causes blistering to the skin. It was, however, the first antiseptic used in operations and saved a lot of lives. Some of its derivatives are not only better germicides, but are also safer to use.

Derivatives of phenol are used in disinfectants

Many household disinfectants are aqueous solutions of phenol derivatives: antiseptics such as TCP, 2,4,6-trichlorophenol and Dettol, 4-chloro-3,5-dimethylphenol and disinfectants such as Jeyes fluid.

phenol — 1
TCP — 23
Dettol — 280

Increasing germ killing power

QUICK QUESTIONS

1 Give the systematic name of TCP and underline the initials that give it its name.

2 Explain why phenol dissolves sparingly in water, but benzene does not dissolve in water.

3 A solution of phenol in water smells of phenol (a disinfectant-type smell). If sodium hydroxide is added, the smell disappears but reappears if concentrated hydrochloric acid is added. Explain why this happens.

4 Phenol burns in air with a sooty flame.
   a What is the soot?
   b Why does phenol, C_6H_5OH burn with a sootier flame than cyclohexanol, $C_6H_{11}OH$?

2.1.1 Introduction

The carbonyl group is shown in the diagram here. **Carbonyl compounds** are either **aldehydes** or **ketones**.

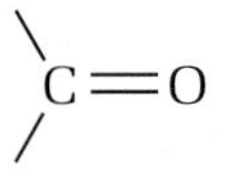

$$\text{\Large $\diagdown$}\!\!C\!=\!O$$

- In aldehydes, the carbon bonded to the oxygen (the carbonyl carbon) has at least one hydrogen atom bonded to it, so the general formula of an aldehyde

is: $\begin{array}{c}R\\ \diagdown\\ C\!=\!O\\ \diagup\\ H\end{array}$ sometimes written as RCHO.

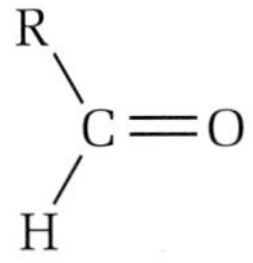

- In ketones, the carbonyl carbon has two organic groups, which we can

represent by R and R' so the formula of a ketone is: $\begin{array}{c}R\\ \diagdown\\ C\!=\!O\\ \diagup\\ R'\end{array}$ sometimes written as RR'C=O

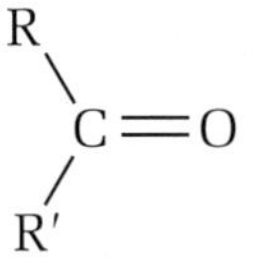

The R-groups in both aldehydes and ketones may be alkyl or aryl.

2.1.2 How to name aldehydes and ketones

Aldehydes are named using the suffix -al. The carbon of the aldehyde functional group is counted as part of the carbon chain of the root. So:

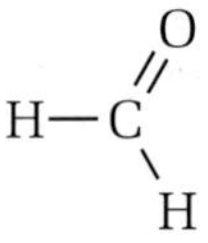

$$H\!-\!\overset{\displaystyle O}{\underset{\displaystyle H}{\overset{\diagup\!\!\diagup}{C}}}\quad \text{or HCHO is methanal and}\quad H\!-\!\overset{\displaystyle H}{\underset{\displaystyle H}{\overset{|}{C}}}\!-\!\overset{\displaystyle O}{\underset{\displaystyle H}{\overset{\diagup\!\!\diagup}{C}}}\quad \text{or CH}_3\text{CHO is ethanal.}$$

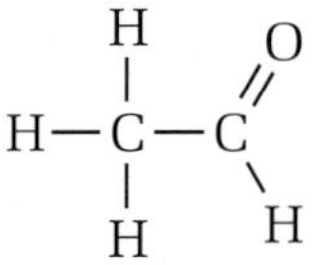

The aldehyde group can only occur at the end of a chain so locants are not needed. When an aldehyde is a substituent on a benzene ring, the suffix -carbaldehyde is used and the carbon is *not* counted as part of the root name.

So: (benzene ring structure with $\overset{O}{\underset{}{\overset{||}{C}}}\diagdown H$) C$_6H_5$CHO, is counted as a derivative of benzene

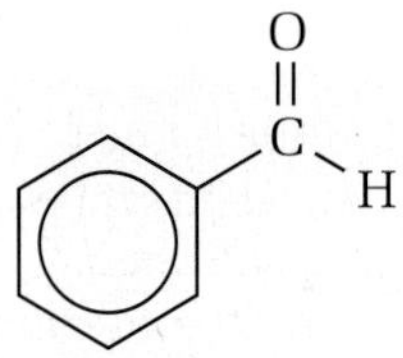

(not of methylbenzene) and is called benzenecarbaldehyde. It is often still called benzaldehyde.

Ketones are named using the suffix -one. In the same way as aldehydes, the carbon atom of the ketone functional group is counted as part of the root. So the simplest ketone:

'Formalin' is a 40% aqueous solution of methanal. It has a fishy, unpleasant smell and is used for preserving biological specimens

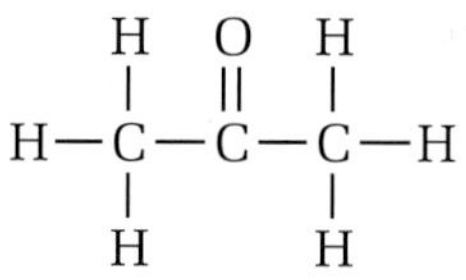

Table 2.1 *A comparison of boiling temperatures*

		M_r	T_b/K
Butane	$CH_3CH_2CH_2CH_3$	60	273
Propanone	CH_3COCH_3	58	359
Propan-1-ol	$CH_3CH_2CH_3OH$	60	370

HINT

You can remind yourself about hydrogen bonding in Topic 3.6 *Essential AS Chemistry for OCR*.

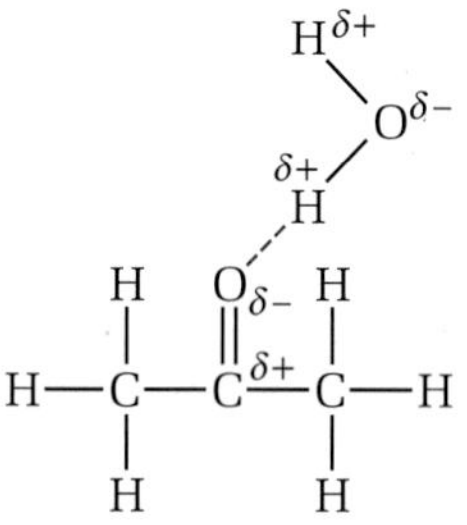

Fig 2.1 *Hydrogen bonding between propanone and water*

DID YOU KNOW?

Propanone is sometimes used for quickly drying laboratory glassware. The wet glassware is rinsed with propanone which then evaporates rapidly as it is much more volatile than water.

Table 2.2 *Comparison of bond strengths*

Bond	Bond enthalpy/kJ mol^{-1}
C=O	740
C=C	612
C—O	358
C—C	347

CH_3COCH_3, is called propanone.

No ketone with fewer than three carbon atoms is possible.

Locants are not needed in propanone or in butanone:

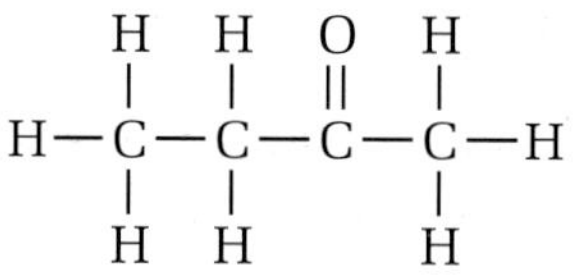

$C_2H_5COCH_3$, because the carbonyl group can only be in one position. With larger numbers of carbon atoms, locants are needed.

2.1.3 Physical properties of carbonyl compounds

The carbonyl group is strongly polar, $C^{\delta+}=O^{\delta-}$, so there are **dipole–dipole forces** between the molecules. These forces mean that:

- Boiling temperatures are higher than those of alkanes with a comparable relative molecular mass but not as high as alcohols, where hydrogen bonding can occur between the molecules, Table 2.1.
- Shorter-chain carbonyl compounds mix completely with water because **hydrogen bonds** form between the oxygen of the carbonyl compound and water, Figure 2.1. As the length of the carbon chain increases carbonyl compounds become less soluble in water.

Methanal, HCHO, is a gas at room temperature. Other short chain aldehydes and ketones are liquids, with characteristic, fairly pleasant smells (propanone is found in many brands of nail varnish remover). Benzenecarbaldehyde smells of almonds and is used to scent soaps and flavour food.

2.1.4 The reactivity of carbonyl compounds

The C=O bond in carbonyl compounds is strong, see Table 2.2 and you might think that the C=O bond would be the least reactive bond. But most reactions of carbonyl compounds involve the C=O bond.

This is because the big difference in electronegativity between carbon and oxygen makes the C=O strongly polar $C^{\delta+}=O^{\delta-}$. So, nucleophilic reagents can attack the $C^{\delta+}$ while electrophiles like H^+ can attack the $O^{\delta-}$. Also, since they contain a double bond, carbonyl compounds are unsaturated and addition reactions are possible. The most typical reactions of the carbonyl group are nucleophilic additions.

QUICK QUESTIONS

1 Name

a
$$CH_3CH_2CCH_2CH_3$$ with O double bonded above the third C

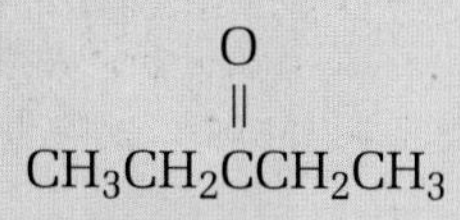

b
$$CH_3CH_2C \text{ with } =O \text{ and } -H$$

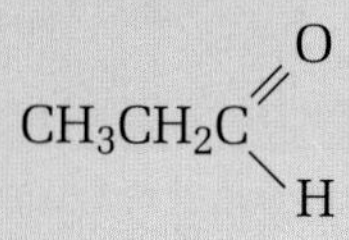

2 Explain why **a** no ketone with fewer that three carbons is possible, **b** a locant is not needed in the name of the ketone butanone, **c** locants are never needed when naming aldehydes.

3 Explain why there are no hydrogen bonds between propanone molecules.

4 Explain why hydrogen bonds can form between propanone and water molecules.

Reactions of carbonyl compounds

As we saw in Section 2.1.3 many of the reactions of carbonyl compounds are nucleophilic addition reactions. They also undergo redox reactions.

2.2.1 Nucleophilic addition reactions

If we represent the nucleophile as :Nu$^-$, the general reaction is:

> **HINT**
>
> Nucleophiles need a lone pair to form bonds with $C^{\delta+}$. Not all nucleophiles are negatively charged (some use the negative end of a dipole to attack $C^{\delta+}$).

The addition of hydrogen cyanide is a good example of a nucleophilic addition:

Addition of hydrogen cyanide
Here the nucleophile is :CN$^-$

With a ketone:

a 2-hydroxynitrile

Or with an aldehyde:

The overall balanced equation for the reaction with an aldehyde is:

$$RCHO + HCN \longrightarrow \underset{\underset{OH}{|}}{\overset{\overset{CN}{|}}{RC}} - H$$

Hydrogen cyanide (in the presence of sodium or potassium cyanide) is used as a source of cyanide ions. (You are unlikely to carry out this reaction in the laboratory because of the toxic nature of the CN$^-$ ion.)

This reaction is useful in organic synthesis because (a) it increases the length of the carbon chain by one carbon and (b) the —CN group can be hydrolysed to a carboxylic acid, —COOH or reduced to an amine —CH$_2$NH$_2$.

2.2.2 Redox reactions

Oxidation
Aldehydes can be oxidised to carboxylic acids.

> **HINT**
>
> The —CN group is called nitrile in an organic compound and cyanide in an inorganic compound.

One oxidising agent commonly used is acidified potassium dichromate often referred to as K$_2$Cr$_2$O$_7$/H$^+$.

Ketones *cannot* be oxidised easily.

HINT

Sodium tetrahydridoborate(III) is sometimes called sodium borohydride.

Reduction

Many reducing agents will reduce both aldehydes and ketones to alcohols. One such reducing agent is sodium tetrahydridoborate(III), $NaBH_4$, in aqueous solution. This generates the nucleophile H^-, the hydride ion.

Note that [H] may also be used to represent reduction in equations. The equation can then be written:

$$\underset{H}{\overset{R}{\diagdown}}C=O \ + \ 2[H] \longrightarrow \underset{H}{\overset{R}{\diagup}}\underset{OH}{\overset{H}{\diagdown}}C$$

Reducing an aldehyde

$$\underset{H}{\overset{R}{\diagdown}}C=O \ + \ H^- \longrightarrow \underset{H}{\overset{R}{\diagup}}\underset{O^-}{\overset{H}{\diagdown}}C \ \xrightarrow[]{(H^+ \text{ from solvent})} \ \underset{H}{\overset{R}{\diagup}}\underset{OH}{\overset{H}{\diagdown}}C$$

(primary alcohol)

Reducing a ketone

$$\underset{R'}{\overset{R}{\diagdown}}C=O \ + \ H^- \longrightarrow \underset{R'}{\overset{R}{\diagup}}\underset{O^-}{\overset{H}{\diagdown}}C \ \xrightarrow[\text{(from solvent)}]{H^+} \ \underset{R'}{\overset{R}{\diagup}}\underset{OH}{\overset{H}{\diagdown}}C$$

secondary alcohol)

These reactions can also be classified as nucleophilic addition reactions, because the H^- ion is a nucleophile and the mechanism is the same.

$$\underset{R'}{\overset{R}{\diagdown}}\overset{:H^-}{\underset{\delta+ \ \ \delta-}{C=O}} \longrightarrow \underset{R'}{\overset{R}{\diagup}}\underset{O: \ H^+}{\overset{H}{\diagdown}}C \ \xrightarrow[]{(H^+ \text{ from solvent})} \ \underset{R'}{\overset{R}{\diagup}}\underset{OH}{\overset{H}{\diagdown}}C$$

(secondary alcohol)

2.2.3 Identifying a carbonyl compound

A variety of instruments is available that will relatively quickly identify most organic compounds, see Chapter 7. A chemical method of identification of an unknown carbonyl compound is:

- React the carbonyl to make a solid (called a solid derivative).
- Measure the melting point of the derivative.
- Check the melting point against a table of melting points of derivatives.

The derivatives are made by using 2,4-dinitrophenylhydrazine, often called Brady's reagent. The carbonyl compound is simply mixed with an acid solution of Brady's reagent in methanol. The derivatives are orange-coloured crystalline solids called 2,4-dinitrophenylhydrazones. These crystals are filtered off and purified by recrystallisation, see Box *Recrystallisation*.

Their melting temperatures are measured, see Figure 2.2. The original carbonyl compound is then identified by referring to tables of melting points of 2,4-dinitrophenylhydrazone derivatives.

RECRYSTALLISATION

Recrystallisation removes impurities from the product. The general idea is that the product dissolves much better in a chosen solvent at a *high* temperature than at room temperature, and precipitates out when the solvent is cooled. Any impurities remain dissolved in the solvent. So, boiling solvent is added to the impure product until just enough has been added to dissolve it all. Any insoluble impurities at this stage are removed by filtering. The mixture is then cooled in an ice bath, when the product will precipitate out and can be removed by filtering.

Brady's reagent can also be used as a test for the presence of a carbonyl compound because orange crystals appear when it is added to either an aldehyde or a ketone.

Distinguishing aldehydes from ketones

Weak oxidising agents can oxidise aldehydes but not ketones. This is the basis of the silver mirror test.

The silver mirror test

Tollen's reagent is a solution of silver nitrate dissolved in aqueous ammonia. It provides Ag^+ ions in an alkaline solution.

- When an aldehyde is warmed with Tollen's reagent, metallic silver is formed. Aldehydes are oxidised to carboxylic acids by Tollen's reagent. The Ag^+ is reduced to metallic silver. A silver mirror will be formed on the inside of the test tube (which has to be spotlessly clean).

$$RCHO + [O] \rightarrow RCOOH \qquad \text{The aldehyde is oxidised.}$$
$$Ag^+ + e^- \rightarrow Ag \qquad \text{The silver is reduced.}$$

- Ketones give no reaction to this test.

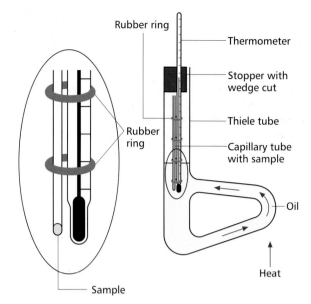

Fig 2.2 *A Thiele tube may be used to measure a melting point*

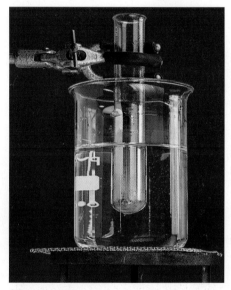

Silver mirror test

QUICK QUESTIONS

1 Which of the following is a nucleophile? H^+, $\text{:}Cl^-$, Cl^-, Cl_2.

2 Sodium tetrahydroborate(III) generates the nucleophile $\text{:}H^-$ and converts carbonyl compounds to alcohols.

   a Would you expect this reagent to reduce

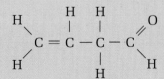

   b Explain your answer.

   c Predict the product when sodium tetrahydroborate(III) reacts with

$$
\begin{array}{c}
H \\
\diagdown \\
C = C - C - C \\
\diagup \quad\quad\quad \diagdown \\
H \quad\quad\quad\quad H
\end{array}
$$

3 Hydrogen with a suitable catalyst will add on to $C=C$ bonds as well as reducing the carbonyl group to an alcohol. Predict the product when hydrogen reacts with the compound in question 2 **c** with a suitable catalyst.

3 Carboxylic acids and esters

3.1 Carboxylic acids

The carboxylic acid **functional group** is

$$-C\underset{O-H}{\overset{O}{\lesseqgtr}}$$

sometimes written

as: —COOH or as —CO$_2$H. This group can only be at the end of a carbon chain.

Carboxylic acids have two functional groups that we have seen before – the

carbonyl group, $\diagdown$C$=$O$\diagup$, found in aldehydes and ketones and the hydroxy

group, —OH, found in alcohols. Having two groups on the same carbon atom changes the properties of each group. The most obvious difference is that the —OH group in carboxylic acids is much more acidic than the —OH group in alcohols.

The most familiar carboxylic acid is ethanoic acid (acetic acid), which is the acid in vinegar.

You have met carboxylic acids before in *Essential AS Chemistry for OCR* where we considered them as the products of vigorous oxidation of primary alcohols.

$$RCH_2OH + 2[O] \rightarrow RCOOH + H_2O$$

3.1.1 How to name carboxylic acids

Carboxylic acids are named using the suffix -oic acid. The carbon atom of the functional group is counted as part of the carbon chain of the root. So:

$$H-C\underset{OH}{\overset{O}{\lesseqgtr}}$$ HCOOH, is methanoic acid; $$H-\underset{\underset{H}{|}}{\overset{\overset{H}{|}}{C}}-C\underset{OH}{\overset{O}{\lesseqgtr}}$$ CH$_3$COOH, is ethanoic

acid and so on.

Where there are substituents or side chains on the carbon chain, they are numbered using **locants**, counting from the carbon of the carboxylic acid as carbon number one. So:

$$H-\underset{\underset{H}{|}}{\overset{\overset{H}{|}}{C}}-\underset{\underset{H}{|}}{\overset{\overset{Br}{|}}{C}}-C\underset{OH}{\overset{O}{\lesseqgtr}}$$ CH$_3$CHBrCOOH, is 2-bromopropanoic acid;

$$H-\underset{\underset{H}{|}}{\overset{\overset{H}{|}}{C}}-\underset{\underset{H}{|}}{\overset{\overset{CH_3}{|}}{C}}-\underset{\underset{H}{|}}{\overset{\overset{H}{|}}{C}}-C\underset{OH}{\overset{O}{\lesseqgtr}}$$ CH$_3$CH(CH$_3$)CH$_2$COOH, is 3-methylbutanoic acid.

When the functional group is attached to a benzene ring, the suffix -carboxylic acid is used and the carbon of the functional group is *not* counted as part of the root. So:

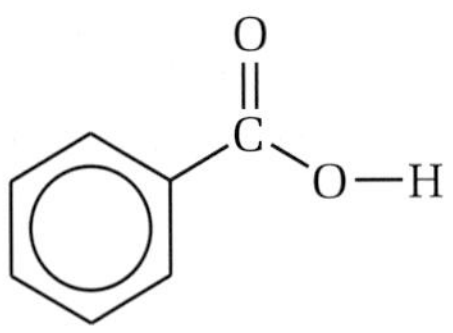

C_6H_5COOH is benzenecarboxylic acid (still often called benzoic acid).

3.1.2 Physical properties of carboxylic acids

The carboxylic acid group can form hydrogen bonds with water molecules, Figure 3.1. For this reason carboxylic acids up to, and including, C_4 (butanoic acid) are completely soluble in water.

The acids also form hydrogen bonds with one another in the solid state. They therefore have much higher melting temperatures than the alkanes of similar relative molecular mass. Ethanoic acid ($M_r = 60$) melts at 290 K while butane ($M_r = 58$) melts at 135 K.

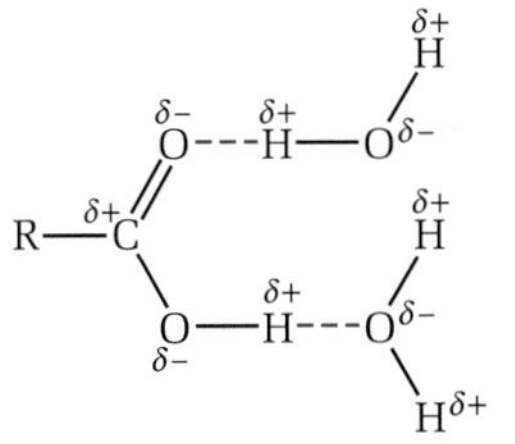

3.1.3 Reactivity of carboxylic acids

The carboxylic acid group is polarised as shown:

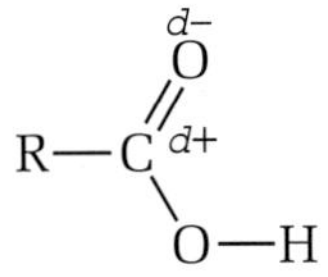

The $C^{\delta+}$ is open to attack from nucleophiles.

The $O^{\delta-}$ may be attacked by positively charged species (like H^+).

But the presence of each group has an effect on the reactivity of the other. The $C^{\delta+}$ tends to attract electrons from the C—OH bond, which makes the $\delta+$ character of this carbon atom less positive. So, it is less easily attacked by nucleophiles than the carbonyl carbon in aldehydes and ketones. The oxygen in the C=O attracts electrons from the O—H bond, weakening this bond. The movement of electrons is shown by the red arrows.

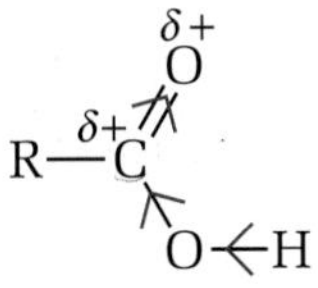

This makes it easier to lose the hydrogen as a H^+ ion.

So the two important features of the chemistry of carboxylic acids are:

- The —OH group easily loses an H^+ ion, so it is a proton donor, making it acidic.
- The carbonyl carbon is susceptible to nucleophilic attack.

Fig 3.1 *A molecule of a carboxylic acid forming hydrogen bonds with water*

DID YOU KNOW?

Pure ethanoic acid is sometimes called 'glacial' ethanoic acid because it may freeze on a cold day. Glacial means ice-like.

The acids have characteristic smells. You will recognise the smell of ethanoic acid as vinegar, while butanoic acid causes the smell of rancid butter.

DID YOU KNOW?

The non-systematic names of hexanoic and octanoic acids are caproic and caprylic acid, respectively, from the same derivation as Capricorn the goat. They are present in goat fat and cause its unpleasant smell.

QUICK QUESTIONS

1 Give the name of

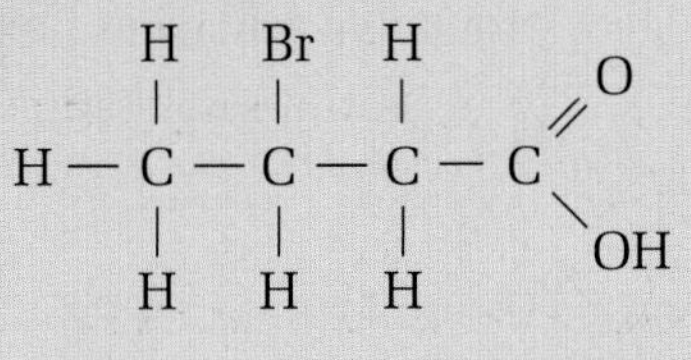

2 Write the displayed formula for 3-chloropropanoic acid.

3 Why is it not necessary to call propanoic acid, 1-propanoic acid?

4 Carboxylic acids, being acidic, will react with the more reactive metals. Give three other reactions that are typical of acids.

3.2.1 Acid reactions

Loss of a proton

If the hydrogen of the —OH group is lost, a negative ion – a carboxylate ion – is left. The negative charge is shared over the whole of the carboxylate group. This delocalisation makes the resulting ion more stable.

a carboxylate ion

In this ion, the negative charge is spread (delocalised) over the atoms shown in red.

The double-headed arrow is used to show that the charge is spread between the two oxygen atoms. It is better represented as:

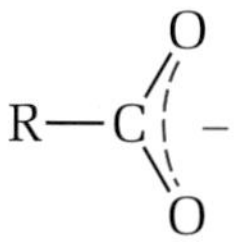

Carboxylic acids are weak acids, so the equilibrium:

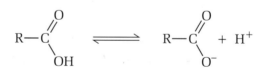

is well over to the left.

Reactions of acids

Carboxylic acids are proton donors and show the typical reactions of acids. They form ionic salts with the more reactive metals, alkalis, metal oxides or metal carbonates in the usual way. The salts that are formed have the general name carboxylates and are named from the particular acid. Methanoic acid gives methanoates, ethanoic acid gives ethanoates, propanoic acid gives propanoates, and so on.

For example,

Ethanoic acid reacts with magnesium:

$$2CH_3COOH(aq) + Mg(s) \rightarrow (CH_3COO)_2Mg(aq) + H_2(g)$$

ethanoic acid   magnesium   magnesium ethanoate   hydrogen

One of the reactions of a carboxylic acid: ethanoic acid reacting with sodium carbonate to produce bubbles of carbon dioxide.

Ethanoic acid reacts with aqueous sodium hydroxide:

$$CH_3COOH(aq) + NaOH(aq) \rightarrow CH_3COONa(aq) + H_2O(l)$$

  ethanoic acid   sodium hydroxide   sodium ethanoate   water

Ethanoic acid reacts with aqueous sodium carbonate:

$$2CH_3COOH(aq) + Na_2CO_3(aq) \rightarrow 2CH_3COONa\ (aq) + H_2O(l) + CO_2(g)$$

  ethanoic acid   sodium carbonate   sodium ethanoate   water   carbon dioxide

3.2.2 Nucleophilic substitution reactions

Formation of esters

Esters have the general formula RCOOR'. The hydrogen from the —OH of the acid is replaced by an alkyl group, so they are acid derivatives.

Carboxylic acids react with alcohols in the presence of a strong acid catalyst to form esters. This is a reversible reaction and forms an equilibrium mixture of reactants and products.

For example:

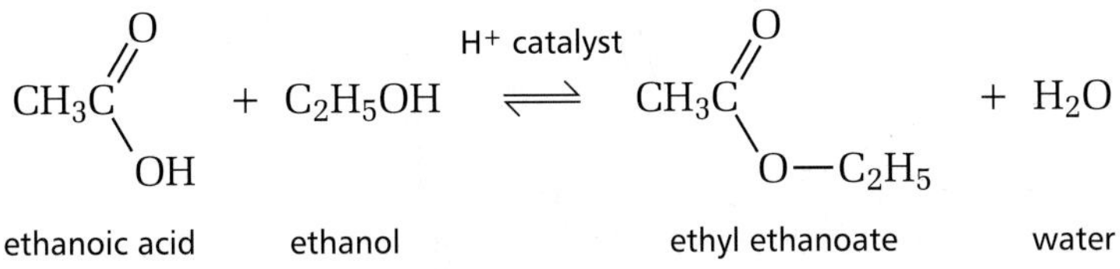

   ethanoic acid   ethanol   ethyl ethanoate   water

Overall, the carboxylic acid and the alcohol react to form an ester and a molecule of water is eliminated. The ester contains the linkage

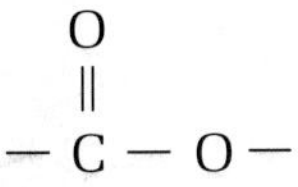

3.2.3 Hydrolysis of esters

The carbonyl carbon atom of an ester has a $\delta+$ charge and is therefore attacked by water acting as a weak nucleophile. The reaction is:

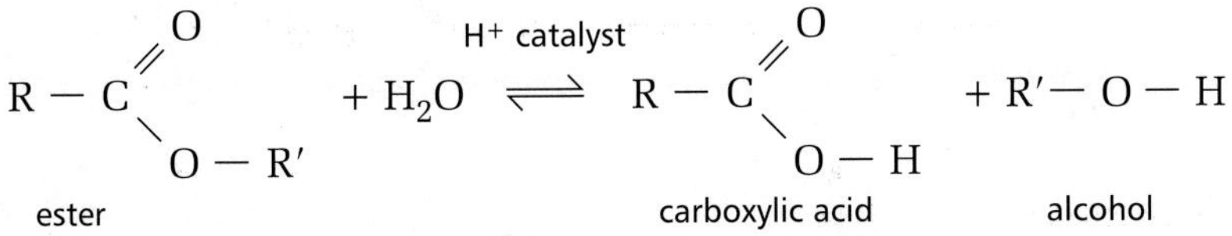

      ester   carboxylic acid   alcohol

The hydrolysis (reaction with water) of esters does not go to completion. It produces an equilibrium mixture containing the ester, water, acid and alcohol. The acid is a catalyst so it affects only the *rate* at which equilibrium is reached, not the composition of the equilibrium mixture.

The reaction takes place at room temperature with a strong acid catalyst. The balanced equation for the acid catalysed hydrolysis of ethyl ethanoate is:

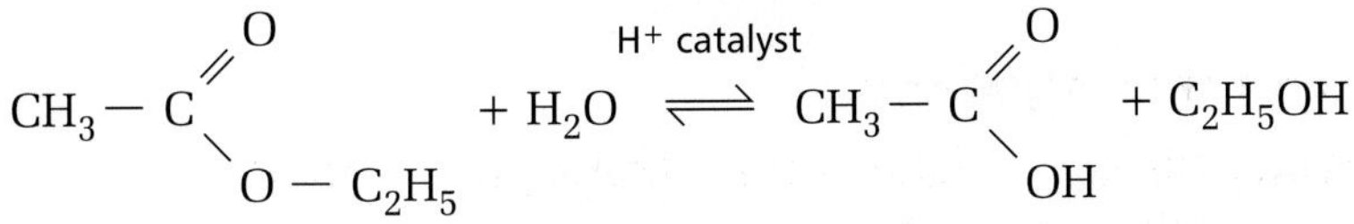

Bases also catalyse hydrolysis of esters. In this case the salt of the acid is produced rather than the acid itself. This removes the acid from the reaction mixture and moves the equilibrium over to the right, so that there is more product in the mixture. For example, methyl ethanoate is hydrolysed to ethanoic acid and methanol, but the ethanoic acid then reacts with the sodium hydroxide to give the salt sodium ethanoate. The reaction takes place at room temperature with a catalyst of sodium hydroxide. The balanced equation for the base catalysed hydrolysis of methyl ethanoate is:

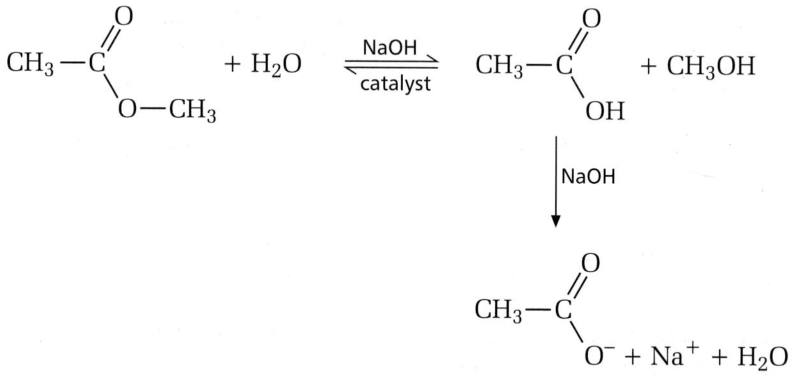

$$CH_3-C\!\!\begin{array}{c} O \\ \| \\ \backslash \\ O-CH_3 \end{array} + H_2O \underset{catalyst}{\overset{NaOH}{\rightleftharpoons}} CH_3-C\!\!\begin{array}{c} O \\ \| \\ \backslash \\ OH \end{array} + CH_3OH$$

$$\downarrow NaOH$$

$$CH_3-C\!\!\begin{array}{c} O \\ \| \\ \backslash \\ O^- + Na^+ + H_2O \end{array}$$

sodium ethanoate

Esters have pleasant smells and tastes and are often found in perfumes and flavourings. For example 3-methylbutyl ethanoate has the smell of pear drops.

QUICK QUESTIONS

1 Write the equation for methanoic acid reacting with lithium.

2 Write the equation for propanoic acid reacting with magnesium oxide.

3 Name the acid and the alcohol that would react together to give the ester methyl ethanoate?

4 Name the acid and the alcohol that would react together to give the ester ethyl methanoate?

5 Methyl ethanoate and ethyl methanoate are a pair of isomers. Explain what this means.

Amines

This chapter is about a group of compounds called amines. Amines can be thought of as derivatives of ammonia in which one or more of the hydrogen atoms in the ammonia molecule have been replaced by organic groups.

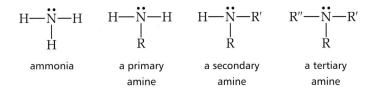

H—N̈—H	H—N̈—H	H—N̈—R′	R″—N̈—R′
H	R	R	R
ammonia	a primary amine	a secondary amine	a tertiary amine

Amines are very reactive compounds so they are useful as intermediates in synthesis – the making of new molecules.

We use the terms **primary**, **secondary** and **tertiary** for amines slightly differently to the way we use them with alcohols. In amines, 1°, 2° and 3° refer to the number of substituents (R-groups) on the *nitrogen* atom. (In alcohols, 1°, 2° and 3° refer to the number of substituents on the *carbon* atom bonded to the —OH group.)

4.1.1 How to name amines

- Primary amines have the general formula RNH_2, where the R can be an alkyl or aryl group. Amines are named using the suffix -amine, for example:

CH_3—NH_2 methylamine

C_2H_5—NH_2 ethylamine

C_6H_5—NH_2 phenylamine (often still called aniline).

- Secondary amines have the general formula RR′NH:

CH_3
  \
    N—H $(CH_3)_2NH$ dimethylamine.
  /
CH_3

- Tertiary amines have the general formula RR′R″N:

C_2H_5
  \
    N—C_2H_5 $(C_2H_5)_3N$ triethylamine.
  /
C_2H_5

Different substituents are written in alphabetical order.

CH_3
  \
    N—H $CH_3(C_3H_7)NH$ methylpropylamine
  /
C_3H_7

HINT

You can remind yourself of lone pairs in *Essential AS Chemistry for OCR* Topic 3.3.

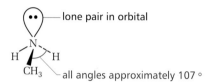

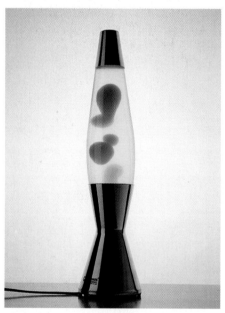

Fig 4.1 *The shape of the methylamine molecule*

HINT

Lower T_b means the molecules are easier to separate.

Phenylamine is used in decorative lava lamps. Phenylamine has almost the same density as water and is not soluble in it. Heat from a bulb at the base of the lamp changes the density enough for the phenylamine to float when hot and sink when cool

4.1.2 The properties of primary amines

Shape

Ammonia is pyramidal with bond angles of approximately 107°. (The angles of a perfect tetrahedron are 109.5°). The difference is caused by the **lone pair**, which repels more than the bonding pairs of electrons in the N—H bonds. Amines keep this basic shape, see Figure 4.1.

Boiling temperatures

Amines are polar:

Primary amines can hydrogen bond to one another *via* their —NH_2 groups (in the same way as alcohols with their —OH groups). However, as nitrogen is less electronegative than oxygen (electronegativities: O = 3.5, N = 3.0), the hydrogen bonds are not as strong as those in alcohols. The boiling temperatures of amines are lower than those of comparable alcohols.

| Methylamine | $M_r = 31$ | CH_3—NH_2 | $T_b = 267\ K$ |
| Methanol | $M_r = 32$ | CH_3—OH | $T_b = 338\ K$ |

Shorter-chain amines such as methylamine and ethylamine are gases at room temperature and those with slightly longer chains are **volatile** liquids. They have fishy smells. Rotting fish and rotting animal flesh smell of di- and triamines, produced when proteins decompose.

Solubility

Primary amines with chain lengths up to about C_4 are very soluble both in water and in alcohols because they form hydrogen bonds with these solvents. Most amines are also soluble in less-polar solvents. Phenylamine is not very soluble in water owing to the benzene ring, which cannot form hydrogen bonds.

4.1.3 The reactivity of amines

Amines have a lone pair of electrons and this is important in the way they react. The lone pair may be used to form a bond with:

- a H^+ ion, when we say the amine is acting as a **base;**
- an electron-deficient carbon atom, when we say the amine is acting as a **nucleophile.**

QUICK QUESTIONS

1 Classify C_2H_5 — $\overset{\overset{\textstyle H}{\textstyle |}}{N}$ — C_3H_7 as primary, secondary or tertiary.

2 Name the compound in question 1.

3 Write the structural formula of trimethylamine.

4 Predict whether dimethylamine will be a solid, liquid or gas at room temperature.

5 Explain your answer to question 4.

The reactions of primary amines

4.2.1 Amines as bases

Amines can accept a proton (an H^+ ion) so they are bases.

$$R\overset{..}{N}H_2 + H^+ \longrightarrow RNH_3^+$$

amine   alkylammonium ion
(or arylammonium ion)

4.2.2 Reaction as bases

Amines react with acids to form salts. For example, ethylamine, a soluble alkyl amine, reacts with dilute hydrochloric acid.

$$C_2H_5\overset{..}{N}H_2 + H^+ + Cl^- \longrightarrow C_2H_5NH_3^+ + Cl^-$$

ethylamine   ethylammonium chloride
(ethylamine hydrochloride)

The products are ionic compounds that will crystallise as the water evaporates.

Phenylamine, an aryl amine, is relatively insoluble, but it will dissolve in excess hydrochloric acid because it forms the soluble ionic salt.

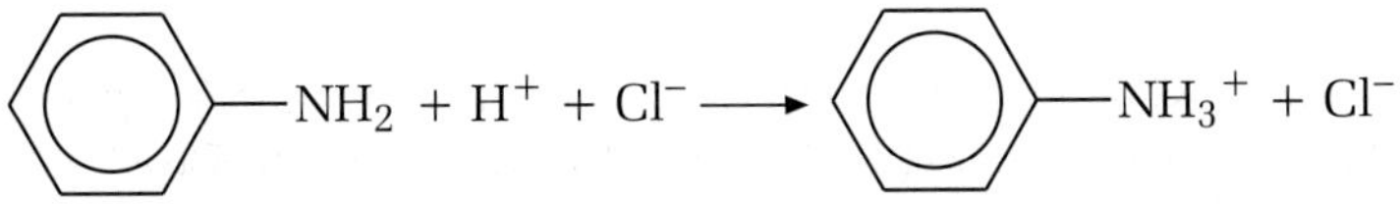

phenylamine   phenylammonium chloride
(phenylamine hydrochloride)
a water-soluble ionic salt

Then, if a strong base, like sodium hydroxide, is added, it removes the proton from the salt and regenerates the insoluble amine.

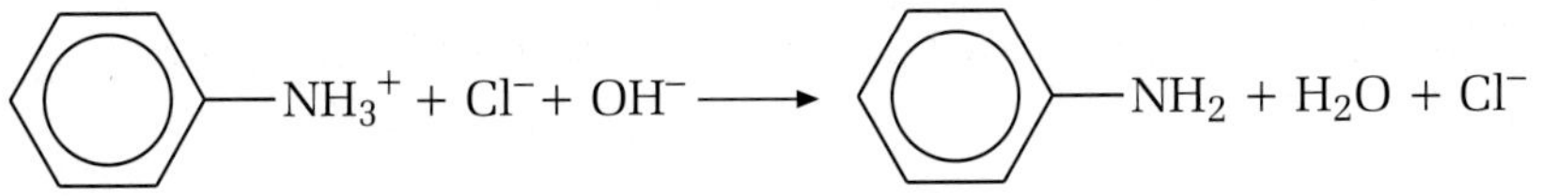

phenylamine

4.2.3 Comparing base strengths

The strength of a base depends on how readily it will accept a proton, H^+. Both ammonia and amines have a lone pair of electrons that attracts a proton.

Alkyl groups *release* electrons away from the alkyl group and towards the nitrogen atom. This is called **the inductive effect** and is shown by an arrow as in Figure 4.2.

The inductive effect of the alkyl group makes the nitrogen atom more negative and therefore more attractive to protons. So, primary alkylamines are stronger bases than ammonia.

Aryl groups *withdraw* electrons from the nitrogen atom because the lone pair of electrons overlaps with the delocalised π-system on the benzene ring as shown

The salts of amines are sometimes named as the hydrochloride of the parent amine.

NOTE
The smell of a solution of an amine disappears when an acid in added, owing to the formation of the ionic, and therefore involatile, salt.

$$R\longrightarrow\overset{..}{N}H_2$$

Fig 4.2 *A primary amine. The arrow shows that R releases electrons. This is called the inductive effect*

for phenylamine.

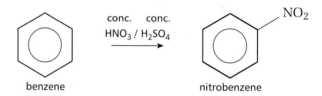

The nitrogen is therefore less attractive to protons, so aryl amines are weaker bases than ammonia.

ethylamine > ammonia > phenylamine

strongest ←——————————————→ weakest

4.2.4 Phenylamine

Phenylamine is the simplest aryl amine. It is the starting point for making many other chemicals and is made in industry using benzene produced from crude oil.

Making phenylamine in the laboratory
Phenylamine can be made in the laboratory from benzene.

Step 1 Benzene is reacted with a mixture of concentrated nitric and sulphuric acids. This produces nitrobenzene.

$$\text{benzene} \quad \xrightarrow[\text{HNO}_3 \,/\, \text{H}_2\text{SO}_4]{\text{conc. \quad conc.}} \quad \text{nitrobenzene} \;\; NO_2$$

Step 2 Nitrobenzene is reduced to phenylamine using tin and hydrochloric acid as the reducing agent.

The tin and hydrochloric acid react to form hydrogen, which reduces the nitrobenzene by removing oxygen atoms of the NO_2 group and replacing them with hydrogen atoms.

$$NO_2 + 6[H] \xrightarrow[\substack{\text{room}\\\text{temperature}}]{\text{Sn / HCl}} NH_2 + 2H_2O$$

This could also be written:

$$C_6H_5NO_2 + 6[H] \rightarrow C_6H_5NH_2 + 2H_2O$$

QUICK QUESTIONS

1 a Write the equation for dimethylamine reacting with hydrochloric acid.

b Give two alternative names for the product.

2 Phenylamine is not very soluble in water. It forms oily drops that float in the water. Predict what you would see if you: **a** add concentrated hydrochloric acid to a mixture of phenylamine and water; **b** then add sodium hydroxide to the resulting solution.

3 Suggest whether dimethylamine will be a weaker or stronger base than ethylamine. Explain your answer.

Azo dyes

Phenylamine and other aryl amines are used to make dyes called azo dyes which contain the linkage —N≡N—. 'Azo' means that the compound contains nitrogen.

4.3.1 Synthesis of an azo dye

This is a two-stage process starting with an aryl amine.

1. Reaction with nitrous acid (nitric(III) acid)
2. Coupling with a phenol.

1. Reaction with nitrous acid

Nitrous acid is unstable and is produced in the reaction vessel by the reaction of sodium nitrite (sodium nitrate(III)) with dilute hydrochloric acid.

$$NaNO_2(aq) + HCl(aq) \rightarrow NaCl(aq) + HNO_2(aq)$$
nitrous acid

Nitrous acid reacts with amines to produce an ionic salt in aqueous solution. The ion R—N$^+$≡N is called a diazonium ion.

$$R{-}NH_2 \xrightarrow{\ HNO_2/HCl\ } R{-}N^+{\equiv}N + Cl^-$$

For example, phenylamine reacts to give a solution of the salt benzenediazonium chloride:

phenylamine   HNO₂ / HCl   benzenediazonium chloride   + Cl⁻

We could write this as a balanced equation:

$$C_6H_5NH_2 + HNO_2 + HCl \rightarrow C_6H_5N_2Cl + 2H_2O$$

(*Alkyl*diazonium ions are unstable and rapidly decompose to give a nitrogen molecule and a carbocation, R$^+$, which will then react rapidly with other compounds in the solution.)

Aryldiazonium ions are more stable than alkyldiazonium ions, providing the solution is kept cold (below about 283 K, 10°C). This is because there is delocalisation of the positive charge onto the aromatic ring – the π-orbitals in the —N$^+$≡N group overlap with the π-system of the benzene ring, see Figure 4.3.

In the solid state they are explosively unstable so they are always used in solution.

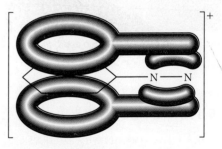

Fig 4.3 *Delocalisation in the benzenediazonium ion*

2. Coupling reactions

Diazonium ions react with phenols. These are electrophilic substitution reactions in which the diazonium ion joins (couples) with the phenol. The reactions are carried out in alkaline conditions at below 10°C.

For example, benzenediazonium chloride reacts with phenol as shown:

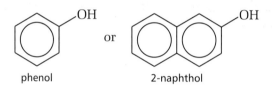

benzenediazonium ion   phenol

4-(phenylazo)phenol
(yellow)

The resulting compounds all contain the unit —N=N— and have *cis-* and *trans-* isomers.

The products of coupling reactions have intense colours. The —N=N— links the π-systems of the two aromatic rings so that the electrons are pooled to give an extended delocalised system, called a conjugated system. Molecules with a conjugated system are often coloured.

4.3.2 Dyeing

Azo dyes are used in the textile industry. The material to be dyed is first soaked in a solution containing a soluble salt of phenol (or a naphthol).

phenol   or   2-naphthol

The cloth is then treated with a diazonium salt and the coupling reaction takes place on the material. So, the dye is formed on the cloth itself, making it resistant to being washed out.

For example:

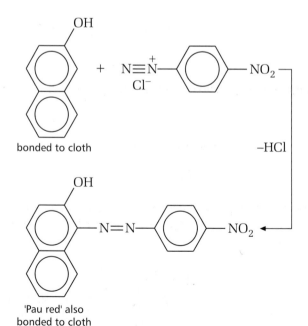

bonded to cloth

–HCl

'Pau red' also
bonded to cloth

Azo dyes are brightly coloured and bond well to fabrics

QUICK QUESTIONS

1 Give two reasons why azo dyes are good dyes.

2 R^+, formed by decomposition of RN_2^+ is best described as:

   a a nucleophile, **b** an electrophile, **c** a free radical, **d** an acid, **e** a base.

3 Explain the term delocalised.

4 Predict the C—N≡N angle in the benzenediazonium ion.

Amino acids

Amino acids have two functional groups – a carboxylic acid and a primary amine. There are about 20 important naturally-occurring amino acids and they are all α-amino acids which means that the amine group is on the carbon next to the —CO_2H group, Figure 4.4.

α-amino acids have the general formula:

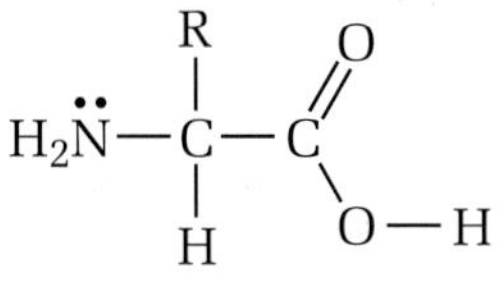

$$H_2\ddot{N}-\underset{\underset{H}{|}}{\overset{\overset{R}{|}}{C}}-C\overset{\displaystyle O}{\underset{\displaystyle O-H}{}}$$

In strongly acidic conditions the lone pair of the $H_2\ddot{N}$— group accepts a proton to form the ion:

$$H-\underset{\underset{H}{|}}{\overset{\overset{H}{|}}{N^+}}-\underset{\underset{H}{|}}{\overset{\overset{R}{|}}{C}}-C\overset{\displaystyle O}{\underset{\displaystyle O-H}{}}$$

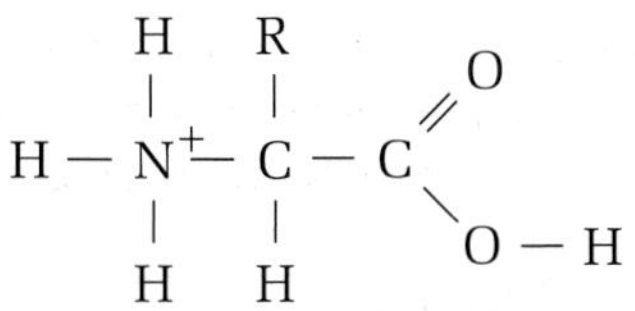

In strongly alkaline solutions the —O—H group loses a proton to form the ion:

$$H_2\ddot{N}-\underset{\underset{H}{|}}{\overset{\overset{R}{|}}{C}}-C\overset{\displaystyle O}{\underset{\displaystyle O^-}{}}$$

$$CH_3-\underset{\underset{H}{|}}{\overset{\overset{NH_2}{|}}{C}}-C\overset{\displaystyle O}{\underset{\displaystyle OH}{}}$$

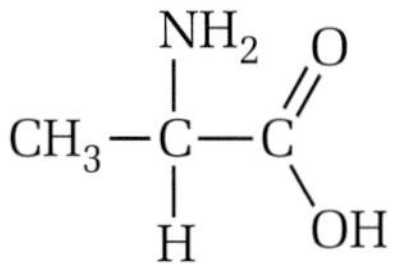

Fig 4.4 α-aminopropanoic acid, also called alanine, written in shorthand as $CH_3CH(NH_2)COOH$

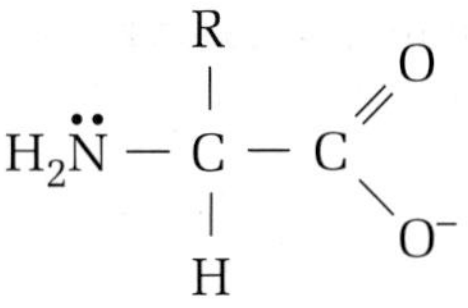

4.4.1 Acid and base properties

Amino acids have both an acidic and a basic functional group:

- The carboxylic acid group has a tendency to lose a proton (act as an acid).

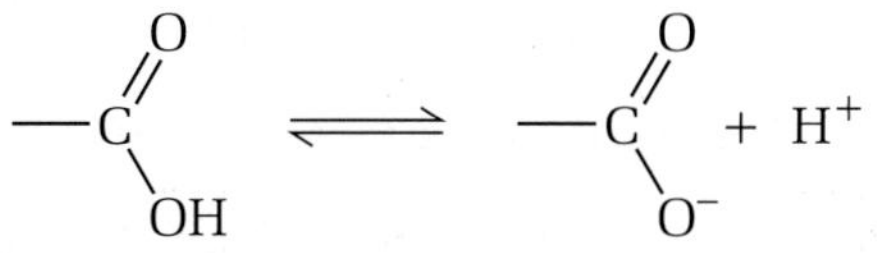

$$-C\overset{\displaystyle O}{\underset{\displaystyle OH}{}} \rightleftharpoons -C\overset{\displaystyle O}{\underset{\displaystyle O^-}{}} + H^+$$

- The amine group has a tendency to accept a proton (act as a base).

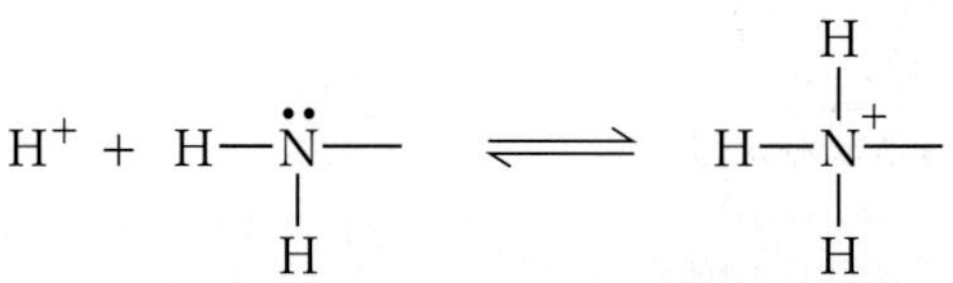

$$H^+ + H-\underset{\underset{H}{|}}{\overset{}{\ddot{N}}}- \rightleftharpoons H-\underset{\underset{H}{|}}{\overset{\overset{H}{|}}{N^+}}-$$

Amino acids exist largely as **zwitterions**. Ions like these have both a permanent positive and negative charge, though the molecule is neutral overall.

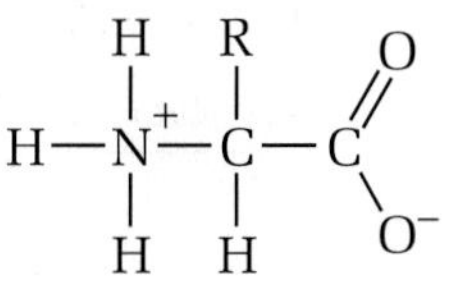

$$H-\underset{\underset{H}{|}}{\overset{\overset{H}{|}}{N^+}}-\underset{\underset{H}{|}}{\overset{\overset{R}{|}}{C}}-C\overset{\displaystyle O}{\underset{\displaystyle O^-}{}}$$

a zwitterion

- The carboxylic acid group has lost a hydrogen ion – we say it is deprotonated.
- The amino group has gained a hydrogen ion – we say it is **protonated**.

Because they are ionic, amino acids have high melting points and dissolve well in water but poorly in non-polar solvents. A typical amino acid is a white solid at room temperature and behaves very much like an ionic salt.

4.4.2 Peptides, polypeptides and proteins

An amide has the functional group $CONH_2$ $-C{\overset{O}{\underset{NH_2}{}}}$ and is formed when ammonia reacts with a carboxylic acid.

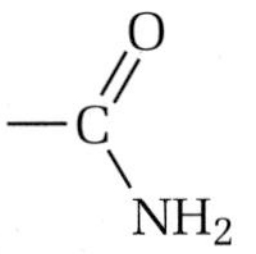

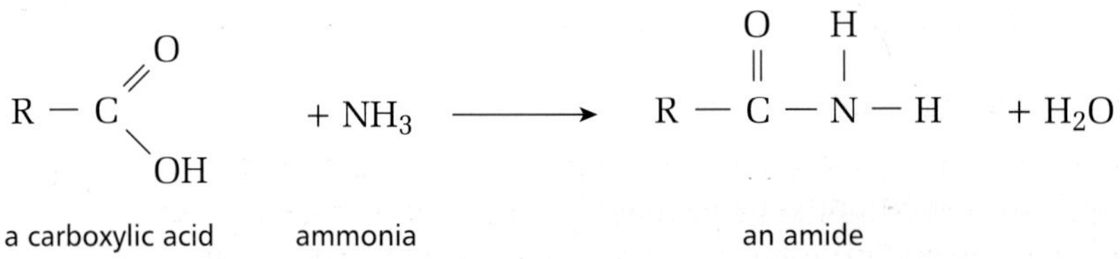

$$R - C{\overset{O}{\underset{OH}{}}} + NH_3 \longrightarrow R - \overset{O}{\overset{\|}{C}} - \overset{H}{\overset{|}{N}} - H + H_2O$$

a carboxylic acid   ammonia   an amide

In a similar way, the amine group of one amino acid can react with the carboxylic acid group of another to form an amide linkage $-CONH-$. This reaction is a nucleophilic substitution in which a molecule of water is eliminated. Compounds formed by the linkage of amino acids are called **peptides** and the amide linkage is often called a peptide linkage.

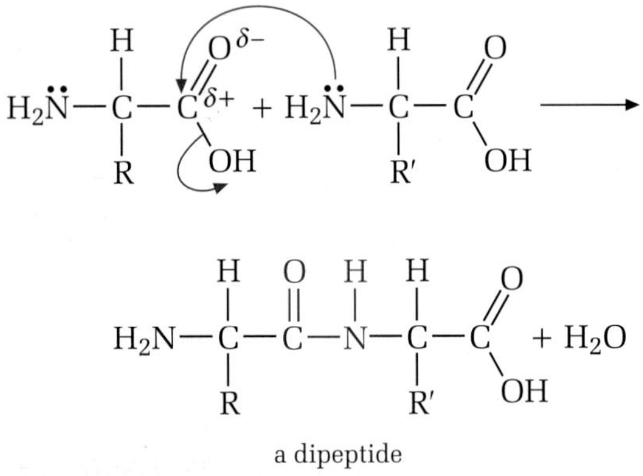

The amide (or peptide) linkage is shown in red in Figure 4.5. The resulting molecule, containing two amino acid units, is called a dimer. A peptide with two amino acids is called a **dipeptide**. The dipeptide still retains $-NH_2$ and $-CO_2H$ groups and so can react further to give tri- and tetrapeptides and so on, see Figure 4.6.

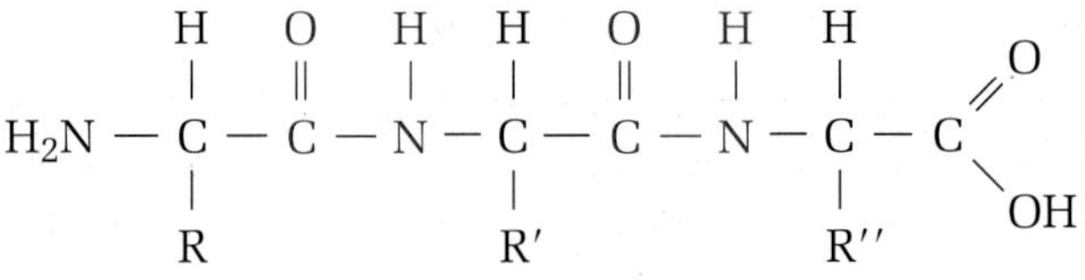

Fig 4.5 Formation of a dipeptide

Fig 4.6 A tripeptide – R, R′ and R″ may be the same or different

Molecules containing up to about 50 amino acids are called **polypeptides** while molecules with more than 50 amino acid units are called proteins. Polypeptides and proteins are condensation polymers, see Topic 6.2. A small molecule, in this case water, is eliminated as each link of the chain forms.

Hydrolysis
When a protein or a peptide is boiled for about 24 hours with hydrochloric acid of concentration 6 mol dm^{-3} it breaks down to a mixture of all the amino acids that made up the original protein or peptide. All the peptide linkages are hydrolysed by the acid, Figure 4.7.

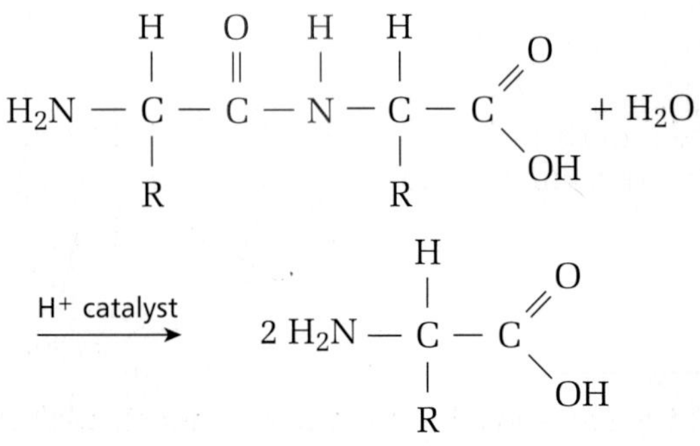

The mixture of amino acids can be separated and identified using paper chromatography.

Fig 4.7 The hydrolysis of the peptide link

QUICK QUESTIONS

1 **a** What are the functional groups in an amino acid?

   b Which group is acidic and which basic?

2 Give the systematic name for .

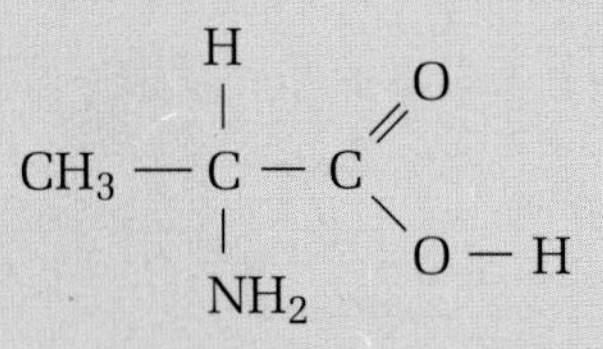

$$CH_3 - \overset{H}{\underset{NH_2}{\overset{|}{C}}} - C{\overset{O}{\underset{O-H}{}}}$$

3 How many amide (peptide) linkages are there in a tripeptide?

4 When an amide (peptide) link is formed

   a what is the nucleophile?

   b which carbon does it attack?

Isomers are compounds with the same molecular formula but a different arrangement of atoms in space. Organic chemistry provides many examples of isomerism. You met structural isomerism in *Essential AS Chemistry for OCR*, Topic 7.4 – the **structural formulae** of the isomers differ. The isomers either have different functional groups or the functional groups are attached to the main chain at different points.

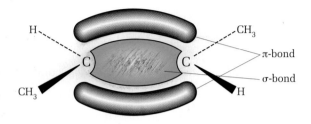

Fig 5.1 *The orbitals forming C=C in trans-but-2-ene*

5.1.1 Stereoisomerism

This is where two (or more) compounds have the same molecular formula and the same functional groups. They differ in the *arrangement* of the groups in space. There are two types: *cis–trans* (or geometrical) isomerism and optical isomerism.

5.1.2 *Cis–trans* isomerism

In organic chemistry, this usually refers to the positions of substituents at either side of a carbon–carbon double bond. Two substituents on either side ao a C=C may either be on the same side of the bond (*cis*) or on opposite sides (*trans*).

cis-1,2-dichloroethene   trans-1,2-dichloroethene

Groups joined by a single bond can rotate around the single bond, but groups joined by a double bond cannot. This is because the p-orbitals that make the π part of the double bond overlap and stop the bond from twisting, see Figure 5.1.

So, *cis* and *trans* isomers are separate compounds and are not easily converted from one to the other.

5.1.3 Optical isomerism

Optical isomers occur when there are four different substituents attached to one carbon atom. This results in two isomers that are mirror images of one another, but are not identical. For example bromochlorofluoromethane exists as two mirror image forms:

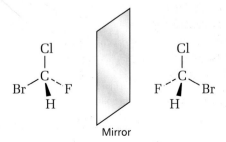

The ball stnd stick models of bromochlorofluoromethane in Figure 5.2 may help you to see that these are not identical.

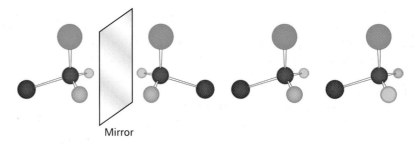

Mirror

Fig 5.2 *Bromochlorofluoromethane has a pair of mirror isomers … … which are not identical*

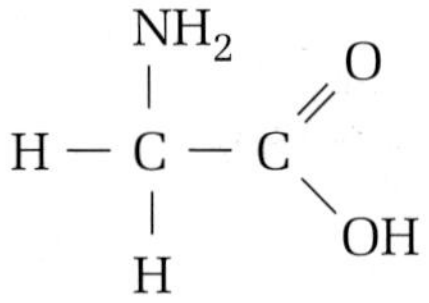

Fig 5.3 *Aminoethanoic acid (glycine)*

glycine

Imagine rotating one of the molecules about the C—Cl bond (pointing upwards) until the two bromine atoms (in red) are in the same position. The positions of the hydrogen (blue) and fluorine atoms (pale green) will not match – you cannot superimpose one molecule onto the other.

This is just like a pair of shoes. A left shoe and a right shoe are mirror images but they are not identical, i.e. they cannot be superimposed – try it!

Pairs of molecules like this are called **optical isomers** because they differ in the way they rotate the plane of polarisation of polarised light (see box *Optical activity*).

5.1.4 Chirality

Optical isomers are said to be **chiral** (pronounced kyral) meaning 'handed' as in left- and right-handed. The carbon bonded to the four different groups is called the **chiral centre** or the **asymmetric carbon atom** and is often indicated by *.
You can easily pick out a chiral molecule because it contains at least one carbon atom that has four different groups attached to it.

- All α-amino acids, except aminoethanoic acid (glycine) (Figure 5.3), the simplest one, have a chiral centre. For example the chiral centre of α-aminopropanoic acid is starred:

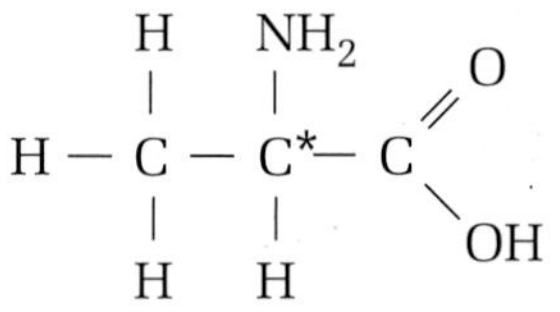

α-aminopropanoic acid

- 2-hydroxyproanoic acid (non-systematic name lactic acid) is also chiral. Although the chiral carbon is bonded to two other carbon atoms, these carbons are part of different groups and you must count the whole group.

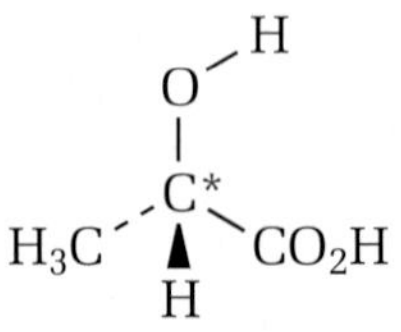

2-hydroxypropanoic acid

Optical isomerism happens because the isomers have three-dimensional structures so they can *only* be shown by three-dimensional representations (or better, by models).

OPTICAL ACTIVITY

Light consists of vibrating electric and magnetic fields. We can think of it as vibrating waves with vibrations occurring in all directions at right angles to the direction of motion of the light wave. If the light passes through a special filter, called a **polaroid** (as in polaroid sunglasses) all the vibrations are cut out except those in one plane, for example the vertical plane.

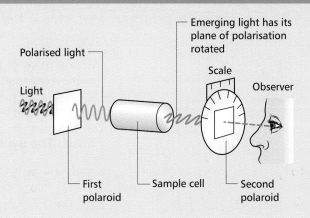

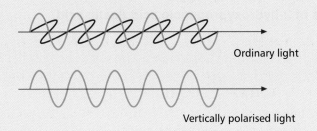

Ordinary light

Vertically polarised light

The light is now vertically polarised and it will be affected differently by different optical isomers of the same substance.

Optical rotation can be measured using a **polarimeter**.

We pass polarised light through two solutions of the same concentration, each containing a different optical isomer of the same substance.

One solution will rotate the plane of polarisation through a particular angle, clockwise. We call this the + isomer.

The other will rotate the plane of polarisation by the same angle, anticlockwise. We call this the − isomer.

QUICK QUESTIONS

1 Which of these molecules can show *cis–trans* isomerism?
a $CH_2\!=\!CH_2$, b $CH_3\!-\!CH_3$, c $RCH\!=\!CH_2$, d $RCH\!=\!CHR$.

2 a Give the name of

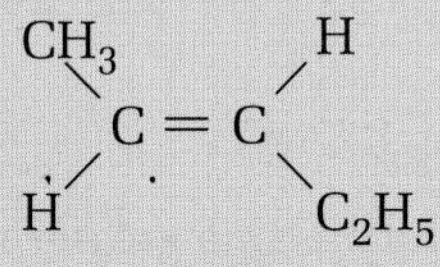

$$\overset{CH_3}{\underset{H}{}}\diagdown C = C \diagup \overset{H}{\underset{C_2H_5}{}}$$

.

b What is the name of its geometrical isomer?

3 Which of the following compounds show optical isomerism?

a
$$\begin{array}{c} H \\ | \\ Cl - C - Br \\ | \\ H \end{array}$$

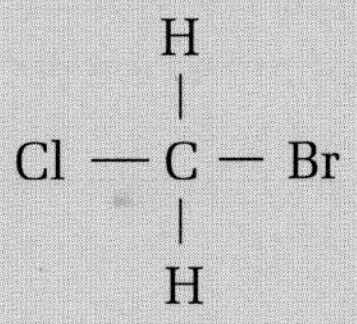

b
$$\begin{array}{c} H \\ | \\ Cl - C - Br \\ | \\ CH_3 \end{array}$$

c
$$\begin{array}{c} H \\ | \\ Cl - C - H \\ | \\ H \end{array}$$

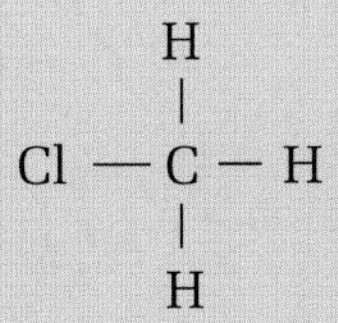

d
$$\begin{array}{c} H \\ | \\ H - C - H \\ | \\ H \end{array}$$

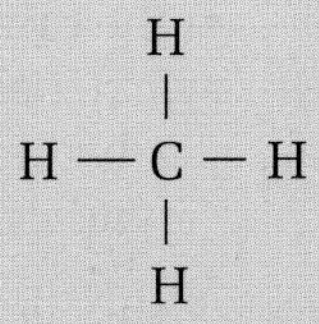

4 Mark the chiral centre on this molecules with a *.

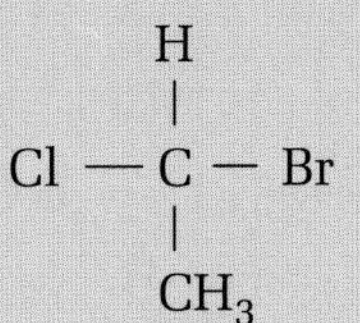

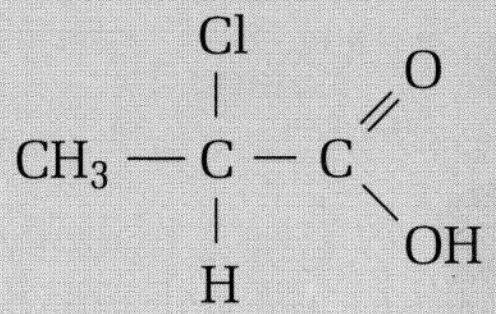

$$CH_3 - \overset{Cl}{\underset{H}{\overset{|}{C}}} - C \diagup^{O}_{\diagdown OH}$$

Synthesis of optically active compounds

A great many of the reactions that are used in synthesis to produce optically active compounds actually produce a 50:50 mixture of two optical isomers. This is called a racemic mixture or racemate (pronounced rass-emm-ate) and is not optically active because the effects of two isomers cancel out.

5.2.1 The synthesis of 2-hydroxypropanoic acid (lactic acid)

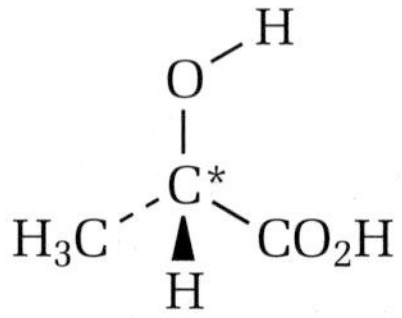

$$H_3C \overset{O-H}{\underset{H}{\overset{|}{\underset{|}{C^*}}}} CO_2H$$

2-hydroxypropanoic acid

2-hydroxypropanoic acid (lactic acid) has a chiral centre, marked by * in the structure above. The synthesis below produces a mixture of optical isomers. It can be made in two stages from ethanal.

Stage 1

Hydrogen cyanide is added across the C=O bond to form 2-hydroxypropanenitrile.

This is a nucleophilic addition reaction in which the nucleophile is the CN^- ion. It takes place as follows:

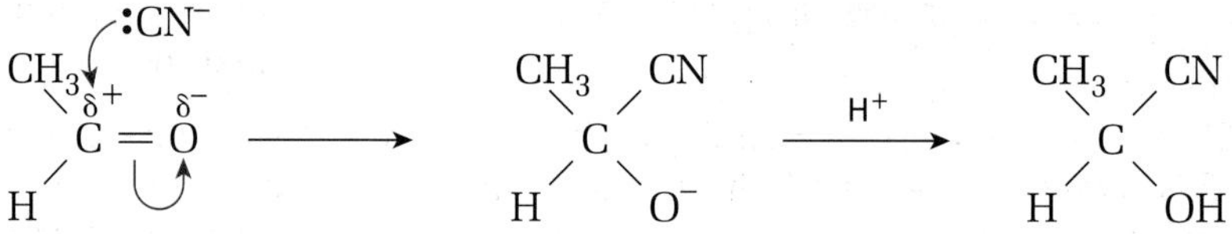

ethanal → → 2-hydroxypropanenitrile

This reaction has two interesting points:

- The carbon chain length of the product is one greater than that of the starting material. We started with *ethan*al (2 C atoms) and ended with 2-hydroxy*propan*enitrile (3 C atoms). This type of reaction is important in synthesis, whenever a carbon chain needs to be lengthened.
- 2-hydroxypropanenitrile has a chiral centre – the starred carbon, which has four different groups (—CH_3, —H, —OH and —CN). The reaction we used does not favour one of these two possible isomers over the other (*i.e.* the —CN group could add on from above or below the CH_3CHO which is flat) so we get a racemic mixture.

Stage 2

The nitrile group is converted into a carboxylic acid group:

The 2-hydroxypropanenitrile is reacted with dilute hydrochloric acid (a hydrolysis reaction):

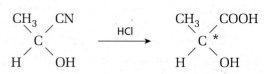

2-hydroxypropanoic acid

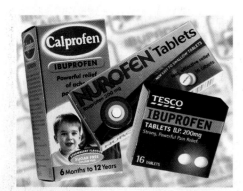

Ibuprofen, Nurofen and Cuprofen. All these drugs contain ibuprofen, which is a mixture of two optical isomers

The balanced equations for the two steps are shown below:

step 1

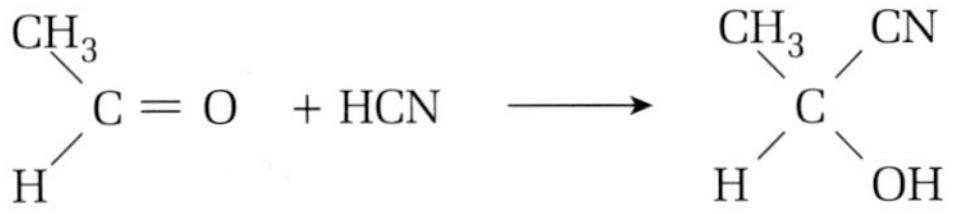

step 2

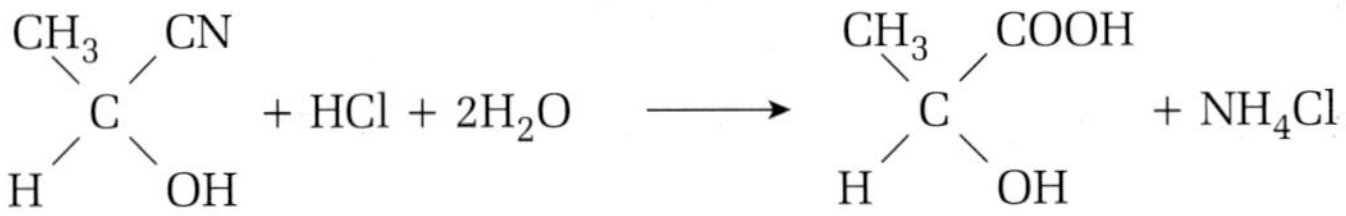

The 2-hydroxypropanoic acid that is produced still has a chiral centre – this has not been affected by the hydrolysis reaction which only involves the —CN group. So we still have a racemic mixture of two optical isomers.

It is often the case that a chemical (with a chiral centre) that is made synthetically, ends up as a mixture of optical isomers, but the same chemical produced naturally in living systems will often be present as only one optical isomer. Amino acids are a good example of this. All naturally occurring amino acids (except aminoethanoic acid, glycine) are chiral, but in every case only one of the isomers is formed in nature. This is because most naturally occurring molecules are made using enzyme catalysts, which only produce one of the possible optical isomers.

5.2.2 Optical isomers in the drug industry

For some purposes, a racemic mixture of two optical isomers will do. For other uses we definitely require one or the other. This is often the case with drugs. Many drugs work by a molecule of the active ingredient fitting an area of a cell (called a receptor) like a piece in a jigsaw puzzle. Because receptors have a three-dimensional structure, only one of a pair of optical isomers will fit. In some cases, one optical isomer is an effective drug and the other is inactive. This is a nuisance – we have three options:

- separate the two isomers – this may be difficult and expensive as optical isomers have very similar properties;
- sell the mixture as a drug – this is wasteful because half of it is inactive;
- design an alternative synthesis of the drug that makes only the required isomer.

The over-the-counter painkiller and anti-inflammatory drug ibuprofen (sold as Nurofen™ and Cuprofen™) is a good example. Its structure is:

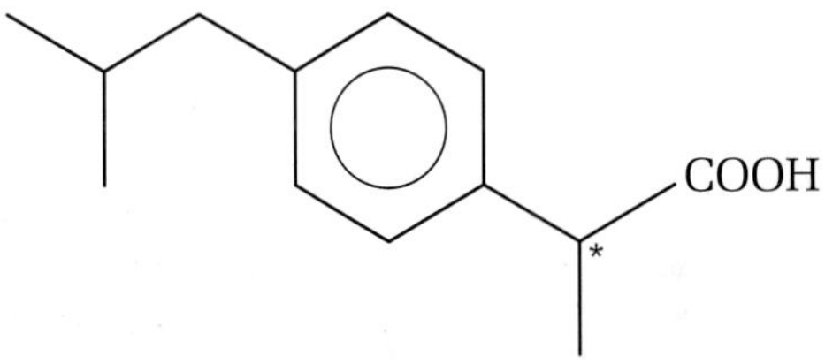

The starred carbon is the chiral centre. Remember that in a skeletal formula, the hydrogen atoms are not drawn, so the starred carbon atom does in fact have four different groups attached.

At present, ibuprofen is made and sold as a racemic mixture.

In other cases one of the optical isomers is an effective drug and the other is toxic or has unacceptable side effects. In this case it is vital that only the effective optical isomer is sold.

QUICK QUESTIONS

1 a What would be the product of the reaction of propanal with hydrogen cyanide followed by reaction with dilute hydrochloric acid?

  b Does this molecule have a chiral centre? Explain your answer.

2 a What would be the product of the reaction of propanone with hydrogen cyanide followed by reaction with dilute hydrochloric acid?

  b Does this molecule have a chiral centre? Explain your answer.

3 Explain why the carbon marked ** in the formula of ibuprofen is not a chiral centre.

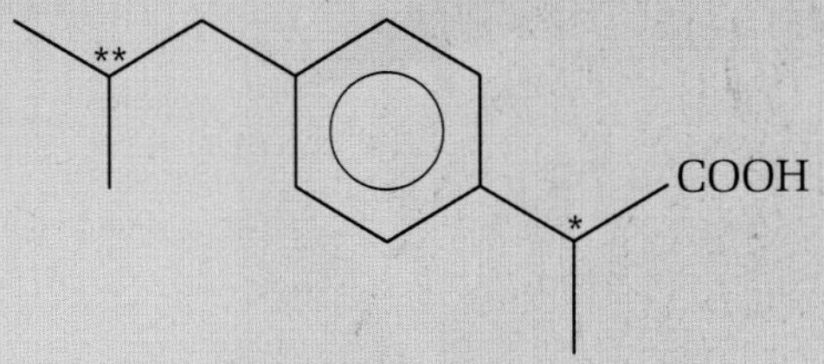

Polymerisation

Different types of polymers

Polymers around us

Polymers are very large molecules that are built up from small molecules, called **monomers**. They occur naturally everywhere – starch, proteins, cellulose and DNA are all polymers. The first completely synthetic polymer was Bakelite, which was patented in 1907. Since then, a large number of different types of polymer have been developed with a range of properties to suit them for very many applications, see photo.

One way of classifying polymers is by the type of reaction by which they are made.

Addition polymers are made from a monomer or monomers with a carbon–carbon double bond.

Condensation polymers are made from a monomer or monomers that have two functional groups. For every bond formed between two monomers, a small molecule, such as water or hydrogen chloride, is expelled.

6.1.1 Addition polymers

We met addition polymers in *Essential AS Chemistry for OCR* Topic 10.4. Addition polymers are made from monomers based on ethene. The monomer has the general formula:

$$\begin{array}{ccc} H & & H \\ & \diagdown\;\diagup & \\ & C=C & \\ & \diagup\;\diagdown & \\ H & & R \end{array}$$

When the monomers polymerise, the double bond opens and the monomers bond together to form a backbone of carbon atoms as shown:

$$\begin{array}{ccccccc} H\ \ \ H & & H\ \ \ H & & H\ \ \ H & \\ \diagdown\;\diagup & & \diagdown\;\diagup & & \diagdown\;\diagup & \\ C=C & + & C=C & + & C=C & \longrightarrow \\ \diagup\;\diagdown & & \diagup\;\diagdown & & \diagup\;\diagdown & \\ H\ \ \ R & & H\ \ \ R & & H\ \ \ R & \end{array}$$

$$-\overset{\displaystyle H}{\underset{\displaystyle H}{C}}-\overset{\displaystyle H}{\underset{\displaystyle R}{C}}-\overset{\displaystyle H}{\underset{\displaystyle H}{C}}-\overset{\displaystyle H}{\underset{\displaystyle R}{C}}-\overset{\displaystyle H}{\underset{\displaystyle H}{C}}-\overset{\displaystyle H}{\underset{\displaystyle R}{C}}-$$

This may also be represented as by equations such as:

$$n\ \ \begin{array}{c} H\ \ \ H \\ |\ \ \ \ | \\ C=C \\ |\ \ \ \ | \\ R\ \ \ R \end{array} \longrightarrow \left[\begin{array}{c} H\ \ \ H \\ |\ \ \ \ | \\ C-C \\ |\ \ \ \ | \\ H\ \ \ R \end{array}\right]_n$$

R— may be an alkyl or an aryl group.

Both the model and the packaging are made from polystyrene (poly(phenylethene))

For example phenylethene polymerises to poly(phenylethene):

H   H   H   H   H   H
 \ /   \ /   \ /
  C = C   +   C = C   +   C = C   + ----
 / \   / \   / \
H   ⬡   H   ⬡   H   ⬡

phenylethene

↓

H H H H H H
| | | | | |
— C — C — C — C — C — C —
| | | | | |
H ⬡ H ⬡ H ⬡

poly(phenylethene)

Phenylethene is sometimes called styrene, which is why poly(phenylethene) is usually called polystyrene.

Table 6.1 gives some examples of addition polymers based on different substituents.

Table 6.1 *Addition polymers made from the monomer* $H_2C=CHR$

R	Name of polymer	Common or trade name
—H	poly(ethene)	polythene (Alkathene)
—CH$_3$	poly(propene)	polypropylene
—Cl	poly(chloroethene)	pvc (polyvinyl chloride)
—C≡N	poly(propenenitrile)	acrylic (Acrilan, Courtelle)
—⬡	poly(phenylethene)	polystyrene

DID YOU KNOW?

The chain length of polythene molecules is typically 5000 ethene units.

6.1.2 The geometry of polymer chains

In an addition polymer there may be different geometrical arrangements of the substituents on the carbon chain. If we imagine the main carbon chain to be in a horizontal plane, (shaded in Figure 6.1) then the substituent (R) may either stick up above or be down below the plane.

- If all the substituents are above the plane, the arrangement is called **isotactic**.
- If substituents are alternately above and below the plane, it is called **syndiotactic.**
- If the substituents are randomly arranged above and below the plane, it is described as **atactic,** see Figure 6.1.

HINT

Remember that hydrocarbon chains are not straight – the C—C—C bond angle is 109.5°

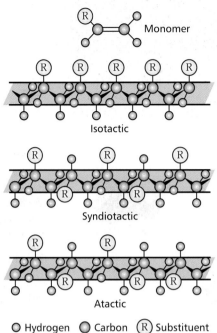

○ Hydrogen ⬤ Carbon Ⓡ Substituent

Fig 6.1 *Geometries of addition polymers*

QUICK QUESTIONS

1 Which of the following monomers could form an addition polymer?

a
H   H
 \   /
  C = C
 /   \
H   H

b
F   F
 \   /
  C = C
 /   \
F   F

c
NH$_2$
|
CH$_3$ — C — COOH
|
H

d
H   H
 \   /
  C = C
 /   \
H   CH$_3$

2 a Draw a section of the polymer formed from the monomer

H   H
 \   /
  C = C
 /   \
H   Cl

, showing six carbon atoms.

   b What is the name of the monomer?

   c What is the systematic name of the polymer?

3

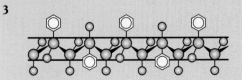

The diagram shows a sections of the polymer poly(phenylethene) drawn so that all the carbon atoms are in the same plane. Is the polymer isotactic, syndiotactic or atactic? Explain your answer.

Condensation polymers

Condensation polymers are normally made from two different monomers each of which has two functional groups. In every condensation reaction a small molecule, often water or hydrogen chloride, is eliminated. Polyesters, polyamides and polypeptides are all condensation polymers. Polyesters, such as Terylene, and polyamides, such as nylons, are commonly used in fibres for clothing.

6.2.1 Polyesters

Esters are formed when carboxylic acids and alcohols react together, see Topic 3.2. This is a condensation reaction because water is eliminated, H from the alcohol and OH from the carboxylic acid. The ester link is in red.

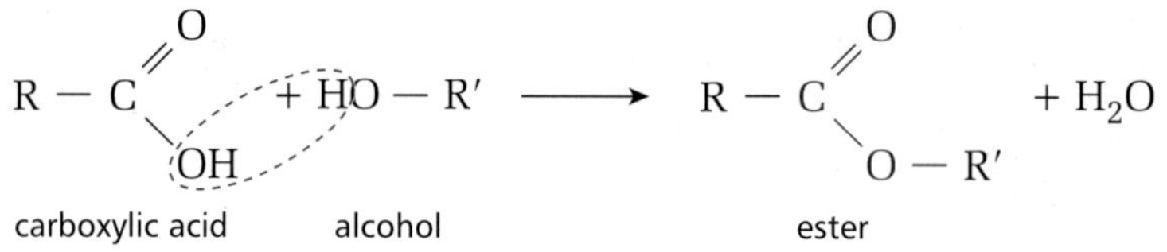

$$R - C \overset{O}{\underset{OH}{\diagdown}} + HO - R' \longrightarrow R - C \overset{O}{\underset{O - R'}{\diagdown}} + H_2O$$

carboxylic acid   alcohol   ester

A *poly*ester has the ester linkage —COO— repeated over and over again. To make a polyester we use diols, which have two —OH groups, and dicarboxylic acids, which have two carboxylic acid, —COOH, groups.

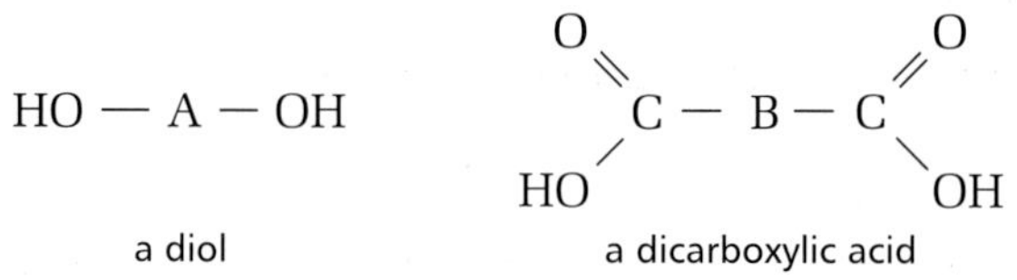

$$HO - A - OH \qquad \overset{O}{\underset{HO}{\diagdown}} C - B - C \overset{O}{\underset{OH}{\diagdown}}$$

a diol   a dicarboxylic acid

The functional groups on the ends of each molecule react to form a chain. For example a diol and a dicarboxylic acid would react together to give a polyester by eliminating molecules of water, Figure 6.2.

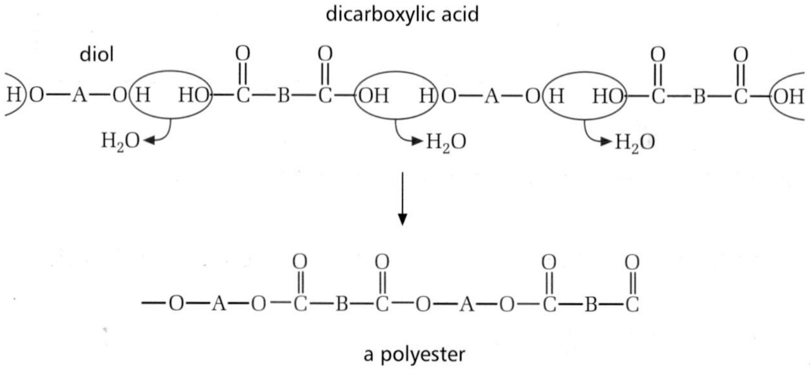

Fig 6.2 *Making a polyester. A and B represent unspecified organic groups, usually* —(CH₂)ₙ

The fibre Terylene is a polyester made from benzene-1,4-dicarboxylic acid and ethane-1,2-diol (Fig. 6.3):

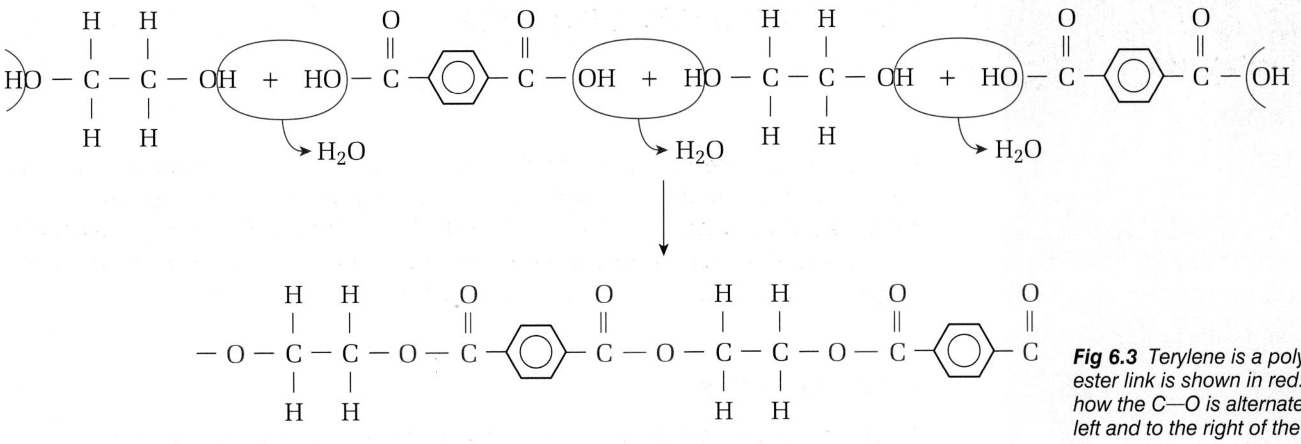

Fig 6.3 *Terylene is a polyester. The ester link is shown in red. Notice how the C—O is alternately to the left and to the right of the C=O*

6.2.2 Polyamides

An amide is formed when an amine and a carboxylic acid react together:

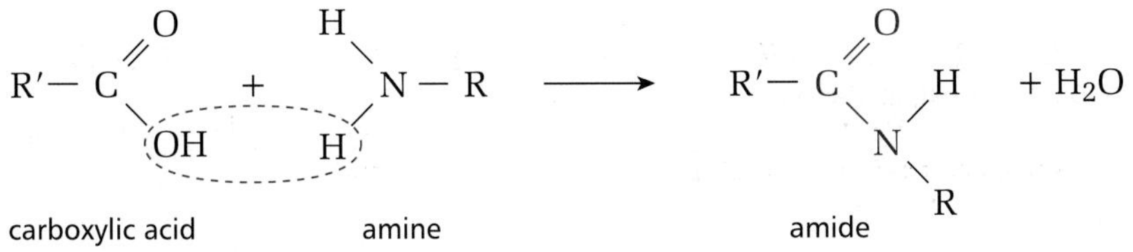

*Poly*amides have the amide linkage —CONH— repeated over and over. To make polyamides from two different monomers a diaminoalkane (which has two amine groups) reacts with a dicarboxylic acid (which has two carboxylic acid groups), Figure 6.4.

Both Nylon and Kevlar are condensation polymers.

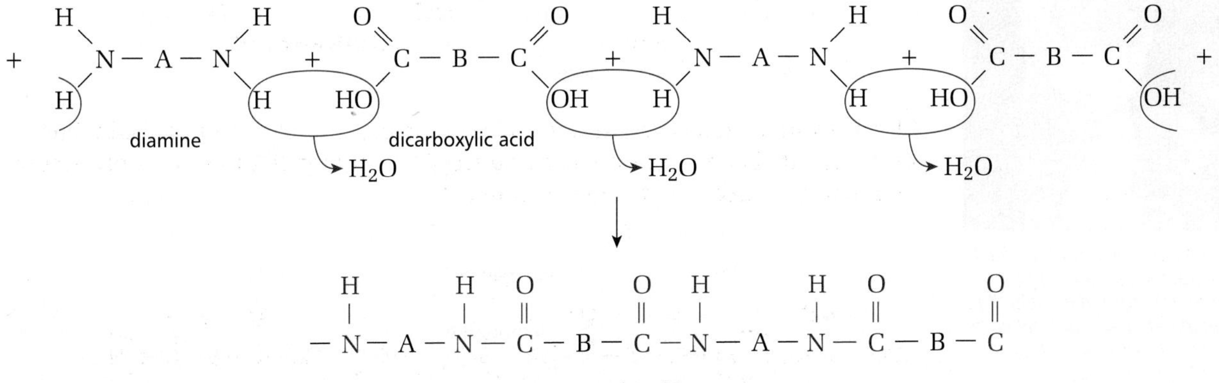

a polyamide

Fig 6.4 *The general equation for making a polyamide, such as Nylon-6,6 or Kevlar*

Nylon

Nylon-6,6 is made from 1,6-diaminohexane and hexane-1,6-dicarboxylic acid:

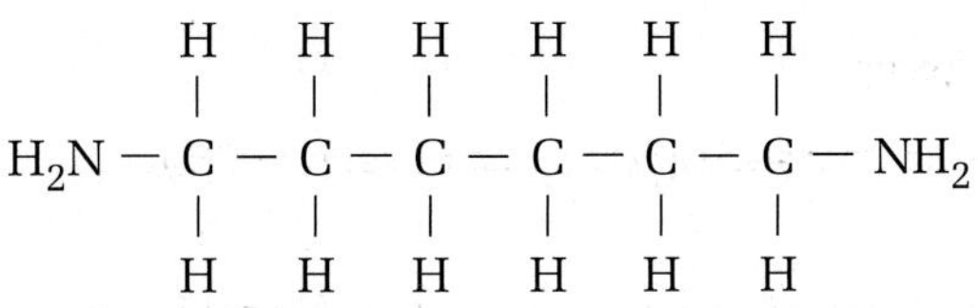

1,6-diaminohexane 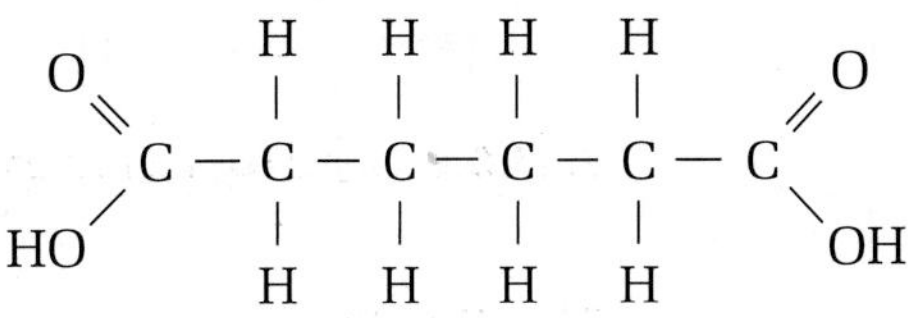   hexane-1,6-dicarboxylic acid

The nylon rope trick. When 1,6-diaminohexane and hexane-1,6-dicarboxylic acid meet, Nylon-6,6 is formed at the interface

Kevlar

Kevlar is made from benzene-1,4-diamine and benzene-1,4-dicarboxylic acid, Figure 6.5.

benzene-1,4-diamine   benzene-1,4-dicarboxylic acid

Kevlar

Fig 6.5 *Kevlar is a polyamide. Because the amide groups are linking rigid benzene rings, Kevlar has very different properties to Nylon*

Polypeptides and proteins

Polypeptides are also polyamides. They may be made from a single amino acid monomer, or many different ones.

As we saw in Topic 4.4, an amino acid has both an amine group and a carboxylic acid group. This means that the amine group of one amino acid can react with the carboxylic acid group of another. A molecule of water is eliminated. The resulting pair of amino acids is called a dipeptide.

amino acids

a dipeptide

Racing drivers wear Kevlar clothing because it is fire resistant and abrasion resistant. For example, it is five times stronger than steel. It is used to replace steel in car tyres, for boat sails, for aircraft wings and in all sorts of protective clothing including bullet-proof vests

The amide linkage is shown in red.

The dipeptide still has —NH_2 and —CO_2H groups so it can go on reacting. Another amino acid can join the chain to form a tripeptide, and so on. A condensation polymer can form, Figure 6.6.

Fig 6.6 *A polypeptide*

Polypeptides are usually made of many different amino acids.

Molecules containing up to about 50 amino acids are called polypeptides while molecules with more than 50 amino acid units are called proteins.

Condensation polymers can be made by simply mixing the monomers, if they are reactive enough.

Notice the difference between a polymer like Nylon, where there are two monomers, one a diamine, $H_2N—X—NH_2$, and one a dicarboxylic acid, $HOOC—Y—COOH$, and a polypeptide, where each amino acid monomer has one $—NH_2$ group and one $—COOH$ group, $H_2N—Z—COOH$. There are twenty naturally occurring varieties of Z.

QUICK QUESTIONS

1 There are a number of different types of Nylon made from two monomers – a dicarboxylic acid and a diamine.

   a The one made from hexane-1,6-dicarboxylic acid and 1,6-diaminohexane is called Nylon-6,6. Suggest where the name comes from.

   b Nylon-6,10 is made from the same dicarboxylic acid as Nylon-6,6. What is the other monomer? Give its name and its formula.

2 Nylon is a polyamide. Explain why proteins and peptides are also called polyamides.

3 Terylene is a polyester made from benzene-1,4-dicarboxylic acid and ethane-1,2-diol. Suggest another diol that would react with this acid to make a different polyester.

Monomers and polymers

Table 6.2 *Addition polymers*

Monomer	Polymer
$CH_2{=}CH_2$	$-[CH{-}CH_2]_n$
$\overset{CH_3}{\underset{\vert}{CH}}{=}CH_2$	$[\overset{CH_3}{\underset{\vert}{CH}}{-}CH_2]_n$
$\overset{Cl}{\underset{\vert}{CH}}{=}CH_2$	$[\overset{Cl}{\underset{\vert}{CH}}{-}CH_2]_n$
$\overset{CN}{\underset{\vert}{CH}}{=}CH_2$	$[\overset{CN}{\underset{\vert}{CH}}{-}CH_2]_n$

Addition polymers and **condensation polymers** start with different types of monomers and their structures are different. This topic is about deducing:

- the structure of the polymer made from given monomers;
- the monomers from given polymer structures.

6.3.1 Predicting the type of polymer from a monomer

The best way to think about this is to remember how the different polymers are formed.

Addition polymers are formed from monomers with carbon–carbon double bonds. There is usually only one monomer (though it is possible to have more), see Table 6.2.

- the monomer (or monomers) have a $C{=}C$ double bond, then an addition polymer will be formed.

Condensation polymers are formed from monomers that have two different functional groups, see Table 6.3.

If there are two monomers, each with two functional groups, or one monomer with two different functional groups then a condensation polymer will be formed.

Table 6.3 *Condensation polymers*

Monomer 1	Monomer 2	Polymer
dicarboxylic acid	diol	
dicarboxylic acid	diamine	
amino acid		

6.3.2 Predicting the type of polymer from a given section of a polymer molecule

You may be asked to identify the type of polymer from a section of the polymer.

- If the main backbone of the polymer is a continuous carbon chain, then it must be an addition polymer. (It may have side chains with functional groups), see Table 6.3.

- If the polymer chain has other atoms in the backbone such as $\overset{O}{\overset{\Vert}{-C-O-}}$

or $\overset{O\ \ H}{\overset{\Vert\ \ \vert}{-C-N-}}$ then it must be a condensation polymer, see Table 6.3.

6.3.3 Finding the repeat unit of a polymer

The **repeat unit** of a polymer is what is says it is – the simplest group of atoms that is repeated throughout the polymer. You find it by starting at any point in a polymer and stopping when the same pattern of atoms begins again, Figure 6.7.

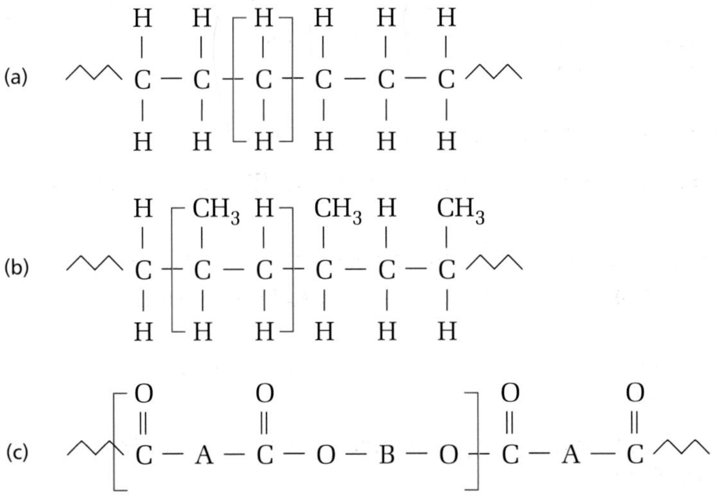

(a)
```
     H   H  ┌H┐ H   H   H
     |   |  │|│ |   |   |
〜〜 C — C ─┤C├─ C — C — C 〜
     |   |  │|│ |   |   |
     H   H  └H┘ H   H   H
```

(b)
```
     H  ┌CH₃ H┐ CH₃ H   CH₃
     |  │|   |│ |   |   |
〜〜 C ─┤C — C├─ C — C — C 〜
     |  │|   |│ |   |   |
     H  └H   H┘ H   H   H
```

(c)
```
   ┌ O      O         ┐ O      O
   │ ‖      ‖         │ ‖      ‖
〜〜┤ C — A — C — O — B — O ├ C — A — C 〜
   └                  ┘
```

Fig 6.7 The repeat unit is inside the red brackets

6.3.4 Finding the monomer(s) of a polymer

Addition polymers

In all addition polymers the **monomer** must contain at least two carbons, so that there can be a carbon–carbon double bond. So, in example (a) in Figure 6.7 the monomer cannot have the same formula as the repeat unit because the repeat unit only has one carbon atom. The monomer is:

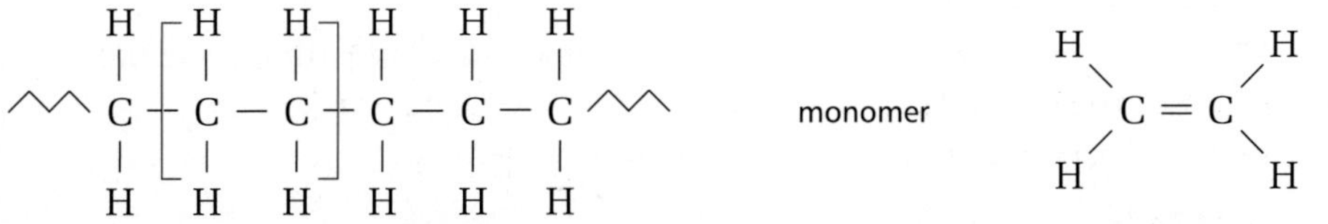

```
     H ┌H   H┐ H   H   H
     | │|   |│ |   |   |
〜〜 C ┤C — C├ C — C — C 〜
     | │|   |│ |   |   |
     H └H   H┘ H   H   H
```
monomer
```
   H                 H
    \               /
     C = C
    /               \
   H                 H
```

In an addition polymer where some, but not all the carbon atoms have substituents, as in example (b) in Figure 6.7, the monomer has the same molecular formula as the repeat unit, (though it must have a double bond).

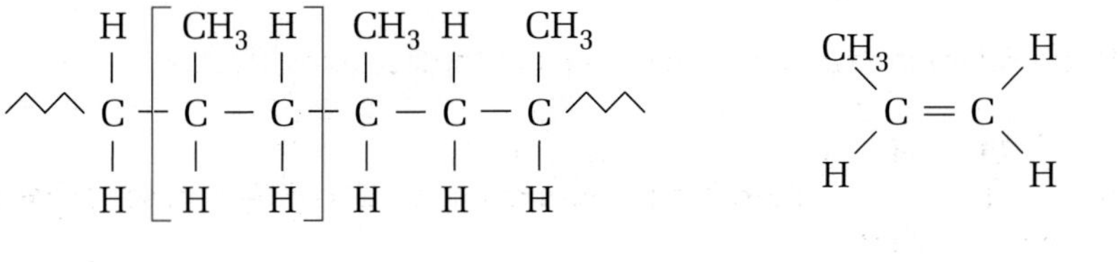

```
     H ┌CH₃ H┐ CH₃ H   CH₃
     | │|   |│ |   |   |
〜〜 C ┤C — C├ C — C — C 〜
     | │|   |│ |   |   |
     H └H   H┘ H   H   H
```
```
  CH₃              H
    \             /
     C = C
    /             \
   H              H
```

Condensation polymers

The best way to work out the monomer(s) in a condensation polymer is to try and recognise the links formed by familiar functional groups, see Table 6.3.

- Start with the repeat unit
- Break the linkage (at the C—O for a polyester or C—N for a polyamide)
- Add back the components of water for each ester or amide link.

For example:

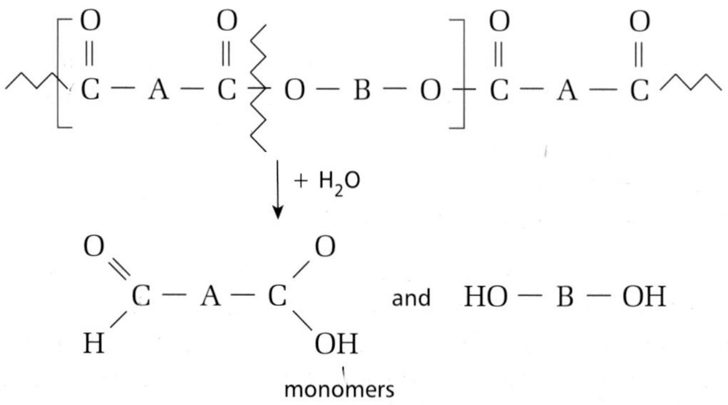

QUICK QUESTIONS

1

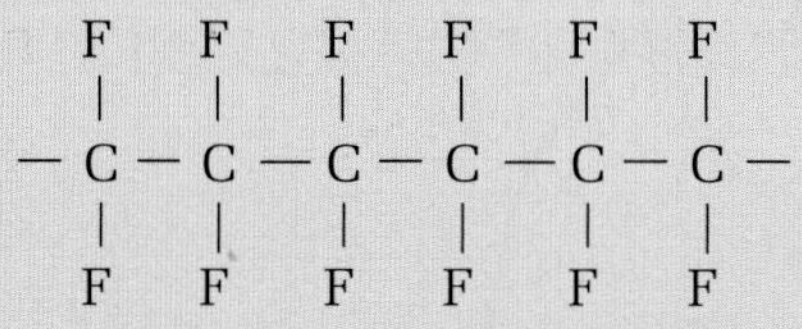

This is a section of the polymer that non-stick pans are coated with the trade name Teflon.

What is **a** the repeat unit, **b** the monomer?

2

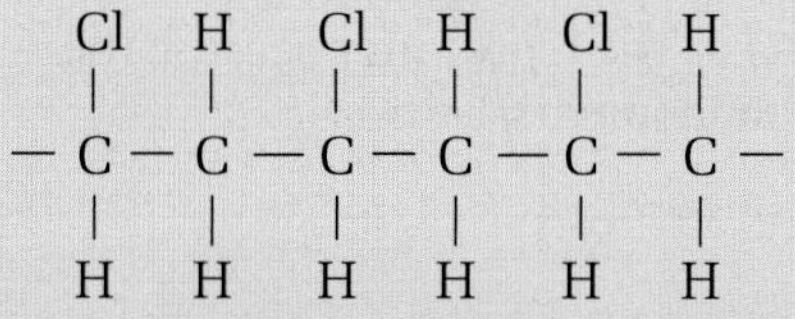

This is a section of the polymer that drainpipes are made from, trade name polyvinylchloride (pvc).

What is **a** the repeat unit, **b** the monomer?

3 Name the type of polymerisation that produces both Teflon and pvc. Explain how you arrived at your answer.

4 What are the linkages called in the two polymers below:

a

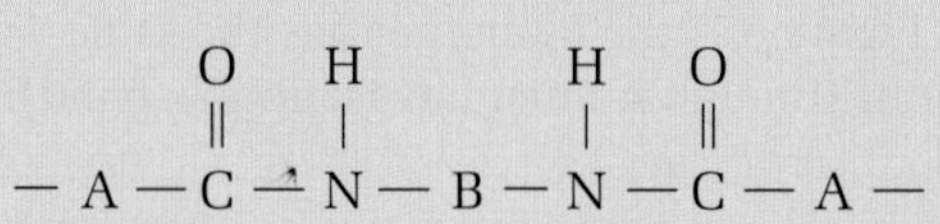

b

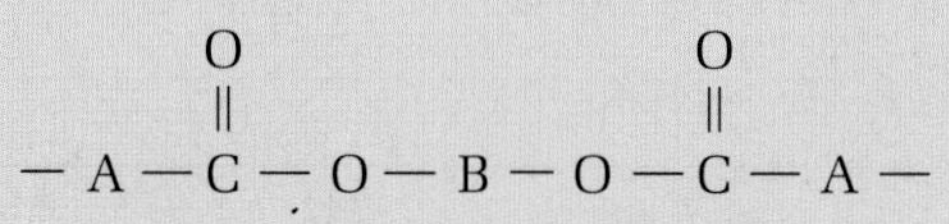

Infra-red (IR) spectroscopy

This module describes three instrumental methods used to help chemists find out the structure of molecules, especially organic ones. These are:

- **Infra-red (IR) spectroscopy**;
- **Mass spectrometry**;
- **Nuclear magnetic resonance, NMR**.

You met infra-red spectroscopy in *Essential AS Chemistry for OCR* (Topic 11.4). An instrument called an infra-red spectrometer is used to identify particular groups of atoms. It produces a graph of transmission against wavenumber for any sample that is investigated. The graph is called a **spectrum** and looks like Figure 7.1. A dip on the spectrum is called a peak.

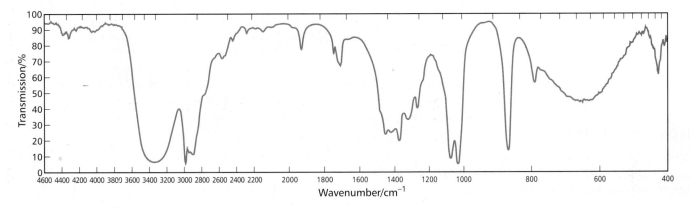

Fig 7.1 *A typical infra-red spectrum of an organic compound*

7.1.1 Identifying groups of atoms

Particular functional groups produce peaks in different areas of the spectrum as summarised in Figure 7.2 and also Table 7.1. The frequencies of peaks are called the wavenumbers and are measured in the units, cm^{-1}.

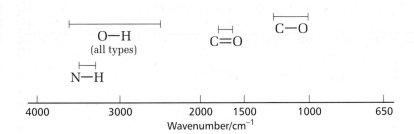

Fig 7.2 *Infra-red absorption of some functional groups*

HOW A SPECTROMETER WORKS

When we shine a beam of infra–red radiation (heat energy) through a sample, the bonds in the sample absorb energy from the beam and vibrate more. But, any particular bond can only absorb radiation that has the same frequency as the natural frequency of the bond. So, the radiation that emerges from the sample (plotted on a graph as **'transmission'**) will be missing the frequencies that correspond to the bonds in the sample. We can identify the groups present from these missing frequencies.

Table 7.1 *Characteristic infra-red absorptions in organic molecules*

Bond	Location	Wavenumber/cm^{-1}
C—O	alcohols, esters	1000–1300
C=O	aldehydes, ketones, carboxylic acids, esters	1680–1750
O—H	hydrogen bonded in carboxylic acids	2500–3300 (broad)
N—H	primary amines	3100–3500
O—H	hydrogen bonded in alcohols, phenols	3230–3550
O—H	free	3580–3670

THE FINGERPRINT REGION

The area of an IR spectrum between 1400 cm^{-1} and 900 cm^{-1} usually has many peaks, caused by the vibrations of the whole molecule. The shape of this region is unique for any particular chemical. So, it can be used to identify the chemical, just as people can be identified by their fingerprints. It is therefore called the fingerprint region

7.1.2 Infra-red spectra of alcohols, R—O—H

The infra-red spectra of alcohols show a peak caused by an O—H vibration at between 3200 cm^{-1} and 3700 cm^{-1}. This vibration is often called a stretching vibration, or just a stretch. The large range is caused by hydrogen bonding between alcohol molecules. Each vibrating hydrogen will be hydrogen-bonded to a varying number of other alcohol molecules, which will tend to slow down the vibration. Therefore the O—H has a range of frequencies depending on how many other molecules are being dragged along. Figure 7.3. shows the infra-red spectrum for propan-2-ol. Notice the broad peak labelled O—H stretch.

Alcohols also show a peak between 1000 and 1300 cm^{-1} caused by a C—O stretching vibration.

QUICK QUESTIONS

Quick questions for this topic will be found at the end of Topic 7.2.

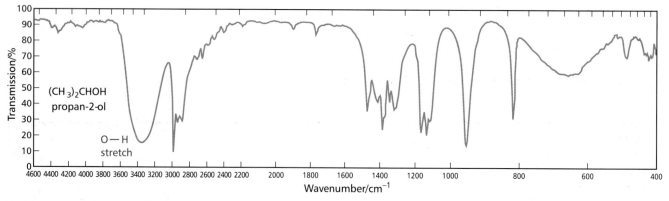

Fig 7.3 *The infra-red spectrum of propan-2-ol*

Infra-red spectra of compounds containing the C=O group

Compounds containing the functional group C=O are called carbonyl compounds. C=O is found in aldehydes, ketones, carboxylic acids, esters and other compounds as well. It has a stretching frequency of between between 1680 and 1750 cm^{-1}.

7.2.1 Infra-red spectra of ketones and aldehydes

The peak produced by ketones and aldehydes is usually strong and sharp. It is between 1680 and 1750 cm^{-1}. The infra-red spectrum of propanone is shown in Figure 7.4. The carbonyl group shows a prominent peak at 1700 cm^{-1} owing to a C=O stretching vibration in the ketone.

$$\underset{\text{a ketone}}{\overset{R}{\underset{R'}{\diagup}}C=O} \qquad \underset{\text{an aldehyde}}{\overset{R}{\underset{H}{\diagup}}C=O}$$

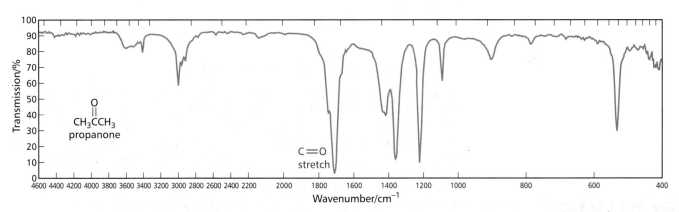

Fig 7.4 *The infra-red spectrum of propanone*

7.2.2 Infra-red spectra of carboxylic acids

$$R-\overset{\overset{\displaystyle O}{\|}}{C}-OH$$

a carboxylic acid

Figure 7.5 shows the infra-red spectrum of ethanoic acid.

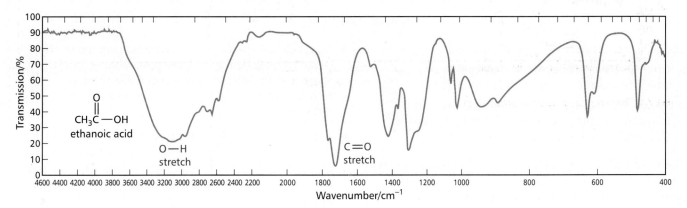

Fig 7.5 *The infra-red spectrum of ethanoic acid*

53

There are three important peaks in the spectrum of ethanoic acid. One is at about 3100 cm^{-1} due to an —OH stretch, another is at about 1700 cm^{-1} due to a C=O stretch and the third is at about 1300 cm^{-1} due to a C—O stretch. We have met both these before in alcohols, in aldehydes and in ketones. The —OH peak is broadened owing to hydrogen bonding, as in the alcohol spectrum, Topic 7.1.

7.2.3 Infra-red spectra of esters

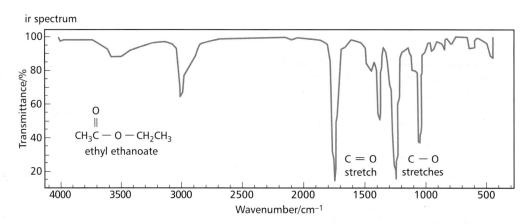

an ester

Esters contain a carbonyl group and therefore show a strong C=O stretching vibration at around 1750 cm^{-1} although this may be shifted up or down by as much as 50 cm^{-1}. The O—H stretch, found in carboxylic acids is now absent, because there is no O—H group in esters, but there are *two* C—O peaks present because there are two C—O bonds. Each C—O bond is slightly different because the atoms around them are slightly different. Figure 7.6 shows the spectrum for ethyl ethanoate.

Fig 7.6 *The infra-red spectrum of ethyl ethanoate*

QUICK QUESTIONS

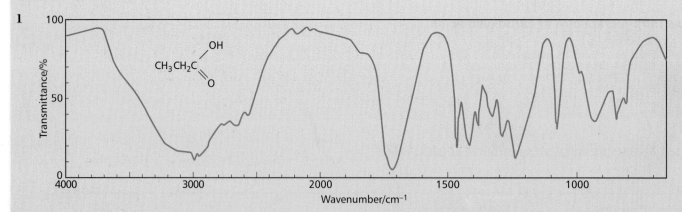

The IR spectrum above is of propanoic acid.

a Give the wavenumber of the peak in the spectrum which shows the presence of (i) C=O, (ii) O—H.

b Which of the above peaks would be present in (i) propanal, (ii) propanone, (iii) methyl propanoate, (iv) propan-1-ol, (v) propan-2-ol?

2 There is an IR peak at about 2800 cm^{-1} caused by C—H stretching vibration. Explain why this is of little use in helping to identify an organic compound.

3 The peak in the IR spectrum representing the O—H stretch appears quite broad in pure ethanol, but it is sharper in the spectrum of ethanol dissolved in tetrachloromethane. Suggest a reason for this difference.

4 The vibrations of the atoms in a chemical bond can be thought of as being like the vibration of a ball hanging on a spring. Using this idea, suggest why the vibration frequency of the O—H bond and that of the N—H bond are so similar.

Mass spectrometry

We saw in *Essential AS Chemistry for OCR*, Topic 1.3, how mass spectrometry is used to measure relative *atomic* masses of isotopes. It is also the main method for finding the relative *molecular* mass of organic compounds. The compound enters the mass spectrometer as a gas or a vapour. It is ionised by an electron gun and the positive ions are accelerated through the instrument as a beam of ionised molecules which are then deflected by a magnetic field. The amount of deflection depends on the mass of the ion. The output is then presented as a graph of abundance (vertical axis) against mass/charge ratio (horizontal axis), but since the charge on the ions is normally 1+, the horizontal axis is effectively relative mass. This graph is called a mass spectrum.

FRAGMENT PEAKS

Although at A2 level you will only need to identify the molecular ion to find the relative molecular mass, the other peaks can give a lot of information about a compound, because they are the peaks that come from the fragments of the molecule. For example, butane and methylpropane are isomers, and have the same relative molecular mass. So, the mass spectra of the isomers will both have the same value for M_r. However, each spectrum has other peaks of different values, because the fragments of the two molecules will be different. Look at the two spectra for butane and methylpropane:

$M_r = 43$, $CH_3CH_2CH_2^+$,

formed when the red bond breaks, and

$M_r = 29$, $CH_3CH_2^+$,

formed when the green bond breaks.

Methylpropane

$$CH_3-\underset{\underset{H}{|}}{\overset{\overset{CH_3}{|}}{C}}-CH_3$$

shows the following main peaks:

$M_r = 58$, molecular ion, $CH_3CH(CH_3)CH_3^+$, and

$M_r = 43$, $CH_3CHCH_3^+$,

formed when any one of the red bonds break.

It is not possible to get a peak of $M_r = 29$ from methylpropane by breaking just one bond.

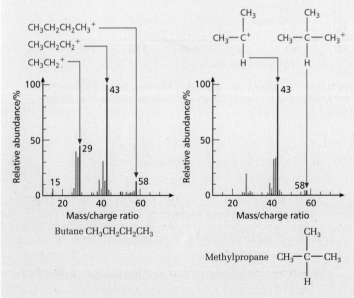

Butane

$$H-\underset{\underset{H}{|}}{\overset{\overset{H}{|}}{C}}-\underset{\underset{H}{|}}{\overset{\overset{H}{|}}{C}}-\underset{\underset{H}{|}}{\overset{\overset{H}{|}}{C}}-\underset{\underset{H}{|}}{\overset{\overset{H}{|}}{C}}-H$$

shows the following main peaks:

$M_r = 58$, molecular ion, $CH_3CH_2CH_2CH_3^+$, and

7.3.1 Measuring relative molecular mass – the molecular ion

Many of the ions will break up – some of their bonds break as they fly through the mass spectrometer. However there are always a few ionised molecules remaining intact to give a peak corresponding to the relative molecular mass, M_r, of the compound. These ionised molecules are called **molecular ions**. The peak furthest to the right of the mass spectrum corresponds to the molecular ion (it has the highest mass). Don't confuse it with the tallest peak in the spectrum, often called the base peak.

Mass spectrometry is the most important technique for measuring M_r.

Figure 7.7 is the mass spectrum of butane. The peak furthest to the right is from the molecular ion, and so $M_r = 58$.

Water boards sample the water from the rivers in their areas to monitor pollutants. The pollutants are separated by chromatography and fed into a mass spectrometer. Each pollutant can be identified from its spectrum – a computer matches its spectrum from a library of spectra of known compounds

$$H-\underset{\underset{H}{|}}{\overset{\overset{H}{|}}{C}}-\underset{\underset{H}{|}}{\overset{\overset{H}{|}}{C}}-\underset{\underset{H}{|}}{\overset{\overset{H}{|}}{C}}-\underset{\underset{H}{|}}{\overset{\overset{H}{|}}{C^+}}-H$$

molecular ion

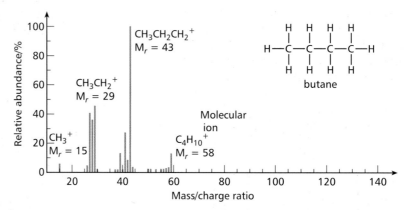

Fig 7.7 *Mass spectrum of butane. The large peaks of masses less than 58 are fragments of the original molecule. The very small peaks are caused by loss of hydrogen atoms from the ions causing the main peaks*

QUICK QUESTIONS

1 Look at the mass spectrum of the organic compound below.

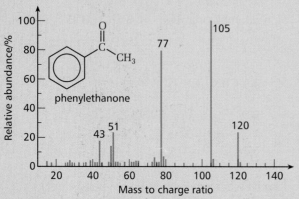

a What is the mass/charge ratio of (i) the molecular ion, (ii) the base peak?

b What is the relative molecular mass of the compound?

c What assumption are you making about the charge on the ion in your answer to **b**?

2 a How are ions formed from molecules in a mass spectrometer?

b What sign of charge do the ions have as a result of this?

3 a Explain how it is possible to get peaks in a mass spectrum that are (i) of smaller mass than the molecular ion (ii) of larger mass than the molecular ion.

b Why are the peaks of larger mass than the molecular ion present in very small abundance?

Nuclear magnetic resonance (NMR) spectroscopy

Nuclear magnetic spectroscopy (NMR) is used particularly in organic chemistry. It is a powerful technique because it helps us find the structure of even quite complex molecules. A magnetic field is applied to a sample along with a source of radio waves.

A BRIEF THEORY OF NMR

Although you will only be examined on *interpreting* the spectra, this background reading may help you to understand how NMR works.

Many nuclei with odd mass numbers, such as ^1H, ^{13}C, ^{15}N, ^{19}F and ^{31}P have the property of *spin* (as do electrons). This gives them a magnetic field like that of a bar magnet.

If magnets are placed in a magnetic field, they will line up parallel to the field as shown in Figure 7.8 (a)

It is also possible that they could line up anti-parallel to the field, as in Figure 7.8 (b), but this has a higher energy, as the magnets have to be forced into that position against the repulsion of the field. They can only stay in the anti-parallel position if undisturbed.

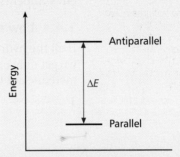

Fig 7.9 *Energy level diagram of the two orientations of bar magnets in a magnetic field*

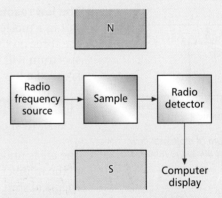

Fig 7.10 *Schematic diagram of an NMR spectrometer*

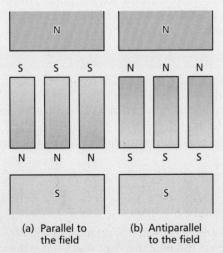

(a) Parallel to the field   (b) Antiparallel to the field

Fig 7.8 (a) and (b) *The two possible orientations of bar magnets in a magnetic field*

The same applies to nuclei with spin, such as hydrogen. They will line up in a magnetic field in the low energy (parallel) position. However if energy just equal to the difference between the two positions (ΔE in Figure 7.9) is supplied, a few nuclei will 'flip' into the higher energy (antiparallel) position. This energy is supplied by a beam of radio waves, see Fig 7.10. NMR instruments in effect measure the energy absorbed by the sample from this beam.

NMR machine

The NMR technique that we are describing is often called *proton* NMR, as it is hydrogen nuclei (protons) that are involved.

Hydrogen atoms in different chemical environments (as part of a CH_2 group or a CHO group, for example) 'feel' the magnetic field differently. This is because all nuclei are **shielded** from the external magnetic field by their electrons. Nuclei with more electrons around them are better shielded. The greater the electron density around a hydrogen atom, the greater magnetic field needed for them to 'flip'. The NMR instrument measures this magnetic field and produces a graph of energy absorbed (from the radio signal) vertically against the magnetic field horizontally. The magnetic field is measured in units called **chemical shift**, symbol δ. The main point about NMR is that hydrogen atoms in different environments will give different chemical shift values.

7.4.1 Low resolution NMR spectra

If all the hydrogen nuclei in an organic compound are in identical environments, we get only one chemical shift value. For example, all the hydrogen atoms in methane are in the same environment:

$$\begin{array}{c} H \\ | \\ H - C - H \\ | \\ H \end{array}$$

methane

The low resolution NMR spectrum of methanol

But in a molecule like methanol, CH_3OH, there are two sorts of hydrogen atoms – the three on the carbon atom, and the one on the oxygen atom. An NMR spectrum will show this – we get the NMR spectrum shown in Figure 7.11. This is called a low resolution spectrum.

> **NOTE**
>
> On NMR spectra chemical shift *decreases* from left to right.

The areas under the peaks (given by the numbers next to them) are proportional to the number of hydrogen atoms of each type – in this case three and one. The chemical shift value at which the peak representing each type of proton appears tells us about its environment.

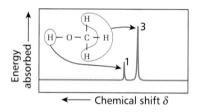

Fig 7.11 *NMR spectrum of methanol (low resolution). Peak areas are in the ratio 1:3*

QUICK QUESTIONS

1 If a bar magnet were pivoted and placed in a magnetic field as shown, describe how it would move in each case (if at all).

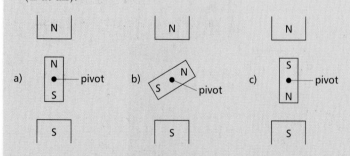

2 **a** Draw the structural formula of ethanol.

  b How many peaks would you expect in the proton NMR spectrum of ethanol?

  c Explain your answer.

  d What would you expect to be the ratio of the heights of the peaks on the low resolution NMR spectrum of ethanol?

  e Explain your answer.

3 Methoxymethane

$$\begin{array}{c} H \quad\quad H \\ | \quad\quad | \\ H - C - O - C - H \\ | \quad\quad | \\ H \quad\quad H \end{array}$$

methoxymethane

is an isomer of ethanol.

  a How many peaks would you expect to find in its low resolution proton NMR spectrum?

  b Explain your answer.

7.5.1 The chemical shift values

The chemical shift, δ, tells us about the chemical environment of hydrogen atoms, see Topic 7.4. In fact each type of hydrogen atom, often referred to as a proton, in any functional group has a particular chemical shift value, see Table 7.3.

Table 7.3 *Chemical shifts for some types of proton in NMR spectrum*

Type of proton	Chemical shift, δ
R—CH$_3$	0.7–1.6
R—CH$_2$—R	1.2–1.4
R$_3$CH	1.6–2.0
—CO—CH$_3$ —CO—CH$_2$—R	2.0–2.9
⬡—CH$_3$ ⬡—CH$_2$—R	2.3–2.7
—O—CH$_3$ —O—CH$_2$—R	3.3–4.3
R—OH	3.5–5.5
⬡—OH	6.5–7.0
⬡—H	7.1–7.7
R—CHO ⬡—CHO	9.5–10
—CO—OH	11.0–11.7

Chemical shifts are typical values and can vary slightly depending on the solvent concentration and substituents.

So, if we are presented with a low resolution spectrum of an organic compound as in Figure 7.12, we can work out its structure.

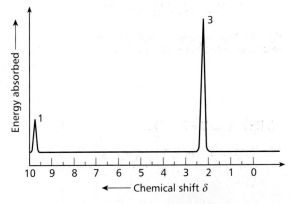

Fig 7.12 *The low resolution NMR spectrum of an organic compound*

- The heights of the peaks (strictly the area underneath them) in an NMR spectrum tell us how many there are of each type of hydrogen atom. These numbers are usually marked on the peaks. Here, there are two types of hydrogen atom, in the ratio 1:3.

- The chemical shift values in Table 7.3 tell us that the single hydrogen at chemical shift 9.7 is the hydrogen from a —CHO (aldehyde) group and the three hydrogens at chemical shift 2.2 are those of a —COCH$_3$ group. (This peak could also be caused by —COCH$_2$R, but since there are three hydrogens it must be —COCH$_3$).

So the compound must be ethanal, CH$_3$CHO, Figure 7.13.

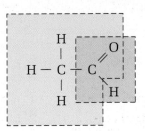

Fig 7.13 *The two groups that make up ethanal*

7.5.2 High resolution NMR – spin–spin splitting

Modern NMR instruments can produce high resolution spectra. A close look at a high resolution spectrum shows that many peaks are split in particular patterns – this is called **spin–spin splitting**. It happens because the NMR signal (from the hydrogen that is causing the peak) is also affected by the magnetic field of the hydrogen atoms on the neighbouring carbon atoms. So, spin–spin splitting gives us information about the neighbouring atoms.

The *n* + 1 rule

- If there is one hydrogen atom on an adjacent carbon this will split the NMR signal of a particular hydrogen into two peaks each of the same height – this is often called a doublet.
- If there are two hydrogen atoms on an adjacent carbon this will split the NMR signal of a particular hydrogen into three peaks in the height ratio 1:2:1 – this is often called a triplet.
- Three adjacent hydrogen atoms will split the NMR signal of a particular hydrogen into four peaks in the height ratio 1:3:3:1 – this is often called a quadruplet or quartet.

This is called the *n* + 1 rule:

> *n* **hydrogens on an adjacent carbon atom will split a peak into *n* + 1 smaller peaks.**

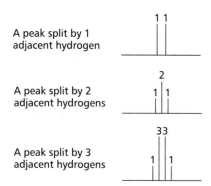

A peak split by 1 adjacent hydrogen

A peak split by 2 adjacent hydrogens

A peak split by 3 adjacent hydrogens

Fig 7.14 *NMR splitting patterns*

Figure 7.14. shows the spin–spin splitting patterns.

7.5.3 Interpreting high resolution NMR spectra

Ethanal

Figure 7.15 shows the *high resolution* NMR spectrum of ethanal.

As we saw from the low resolution spectrum, here are two types of hydrogen:

- There is a single hydrogen of chemical shift 9.7. This is the hydrogen of a —CHO group. This peak is split into 4 (height ratios 1:3:3:1) by the three hydrogens of the adjacent —CH₃ group.
- At chemical shift 2.2 are three hydrogens of a —CH₃ group. This peak is split into two (height ratios 1:1) by the one hydrogen of the adjacent —CHO group.

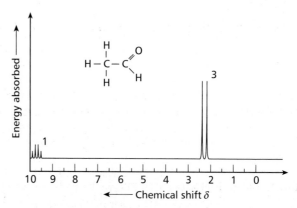

Fig 7.15 *The high resolution NMR spectrum of ethanal, CH₃CHO*

Propanoic acid

Figure 7.16 shows the high resolution NMR spectrum of propanoic acid.

It is useful to make a table of the peaks by reference to Table 7.3.

Chemical shift, δ	Type of hydrogen	Number of hydrogens
11.7	—COO**H**	1
2.4	—CO**CH₂**R or —CO**CH₃**	2
1.1	R**CH₃**	3

From the chemical shift value alone, the peak at 2.4 could be either of the groups shown. However the fact that there are just two hydrogens means that it must correspond to —CO**CH₂**R.

Looking at the spin–spin splitting:

The peak at 11.7 is not split. This is because the adjacent carbon has no hydrogens bonded to it, —COOH

The peak at 2.4 is split into four. This indicates that the adjacent carbon has three hydrogens bonded to it. So, the R in —CO**CH₂**R must be —CH₃.

The peak at 1.1 is split into three. This indicates that the adjacent carbon has two hydrogens bonded to it. So, the R in R**CH₃** must be —CH₂.

So, if we put these groups together we make propanoic acid.

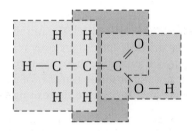

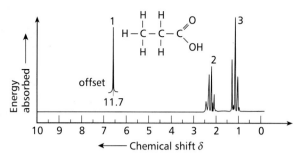

Fig 7.16 *The high resolution NMR spectrum of propanoic acid, CH₃CH₂COOH. Note how the peak at $\delta = 11.7$, which is off the normal scale, is presented*

QUICK QUESTIONS

1 This question is about the isomers propan-1-ol and propan-2-ol.

What is meant by the term isomer?

2 Write down the formulae of propan-1-ol and propan-2-ol and mark each of the hydrogen atoms A, B etc. to show which are in different environments.

a How many different environments are there for the hydrogen atoms in
(i) propan-1-ol, (ii) propan-2-ol?

b How many hydrogen atoms in each of the different environments, A, B etc, are there in
(i) propan-1-ol, (ii) propan-2-ol?

c Predict the chemical shift for each environment in
(i) propan-1-ol, (ii) propan-2-ol?

More examples of interpreting and predicting NMR spectra

7.6.1 Interpreting NMR spectra

Propanone

The NMR spectrum of propanone has just one peak, see Figure 7.17. This means that all the hydrogen atoms in the molecule are identical. The chemical shift value of 2.1 indicates that this corresponds to —$COCH_3$ or —$COCH_2R$.

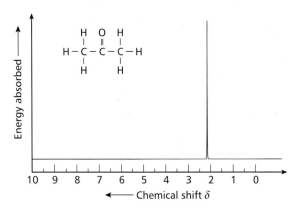

Fig 7.17 *The high resolution NMR spectrum of propanone*

> **NOTE**
>
> There are six hydrogen atoms in a molecule of propanone, all of which are in an identical environment. However, we cannot tell the number from the spectrum because the peak areas in NMR spectra are only relative. This means that we can tell that in methanol, for example, there are three times as many —CH_3 peaks as —OH peaks but not the absolute number of each type of hydrogen (the absolute numbers could be 2 and 6, for example).

7.6.2 Predicting NMR spectra

Chemists making new compounds may predict the spectrum of a compound they are making and compare their prediction with that of the compound they actually produce to check that their reaction has gone as intended.

Ethyl ethanoate

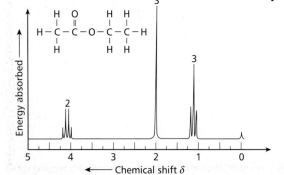

Fig 7.18 *The high resolution spectrum of ethyl ethanoate*

There are three sets of hydrogen atoms in different environments. The values of chemical shift are predicted using Table 7.3.

We can predict the spectrum shown in Figure 7.18, by dividing up the molecule as shown:

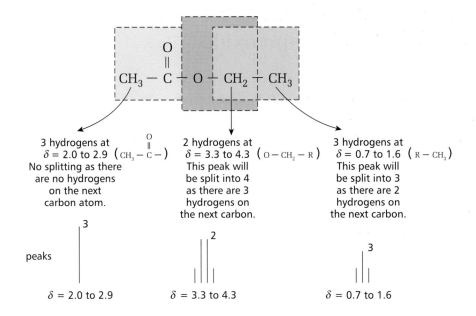

3 hydrogens at $\delta = 2.0$ to 2.9 $\left(\underset{\text{CH}_3-\overset{\displaystyle \text{O}}{\underset{\displaystyle \|}{\text{C}}}-\right)$ No splitting as there are no hydrogens on the next carbon atom.

2 hydrogens at $\delta = 3.3$ to 4.3 $\left(\text{O}-\text{CH}_2-\text{R}\right)$ This peak will be split into 4 as there are 3 hydrogens on the next carbon.

3 hydrogens at $\delta = 0.7$ to 1.6 $\left(\text{R}-\text{CH}_3\right)$ This peak will be split into 3 as there are 2 hydrogens on the next carbon.

peaks

$\delta = 2.0$ to 2.9 $\delta = 3.3$ to 4.3 $\delta = 0.7$ to 1.6

QUICK QUESTIONS

1 The NMR spectra below are of two compounds, A and B,

one of which is ethanol $\text{H}-\overset{\displaystyle \text{H}}{\underset{\displaystyle \text{H}}{\text{C}}}-\overset{\displaystyle \text{H}}{\underset{\displaystyle \text{H}}{\text{C}}}-\text{O}-\text{H}$ and the

other methoxymethane $\text{H}-\overset{\displaystyle \text{H}}{\underset{\displaystyle \text{H}}{\text{C}}}-\text{O}-\overset{\displaystyle \text{H}}{\underset{\displaystyle \text{H}}{\text{C}}}-\text{H}$.

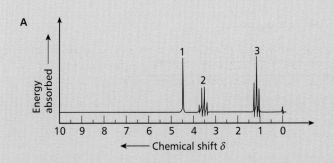

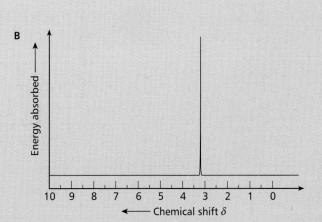

a Work out which spectrum represents which compound.

b Say what type of hydrogen each peak represents.

c How many of each type of hydrogen are there?

2 Predict the NMR spectrum of methyl ethanoate, $\text{CH}_3\text{COOCH}_3$, using the same procedure as for ethyl ethanoate above.

63

7.7

The use of deuterium oxide

Deuterium, symbol D, is an isotope of hydrogen which has a proton *and a neutron* in the nucleus. It behaves chemically just like hydrogen but its nucleus has no spin and it therefore gives no NMR signal. Deuterium oxide, D_2O is water in which the hydrogen atoms are replaced by deuterium.

Hydrogen atoms that are bonded to carbon atoms in organic compounds are not affected by deuterium oxide. Hydrogen atoms that are bonded to electronegative atoms such as those in —OH or —NH groups will react with deuterium oxide. These hydrogen atoms will exchange places with the deuterium. This is called deuterium exchange, e.g.

$$ROH + D_2O \rightarrow ROD + DHO$$

All we need to do is shake the organic compound with deuterium oxide.

If we do this, the NMR signal due to the —OH or —NH hydrogen disappears. This can help us to analyse an NMR spectrum by confirming which peak is caused by hydrogen atoms in —OH or —NH groups.

Any spin–spin splitting caused by these hydrogens will also disappear from the spectrum.

Figure 7.19 (a) shows the NMR spectrum of pure methanol and Figure 7.19 (b) shows the spectrum of methanol after deuterium exchange. Notice that the —OH signal at chemical shift 4.1 has disappeared, as has the splitting of the —CH_3 peak caused by the hydrogen of the —OH group.

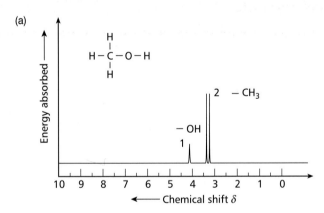

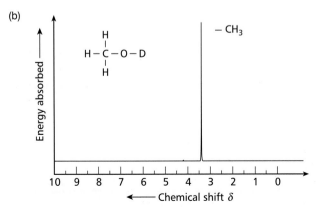

Fig 7.19 *The NMR spectra of methanol (a) before and (b) after deuterium exchange*

QUICK QUESTIONS

1 The NMR spectrum of ethanoic acid, H—C—C, is shown below.

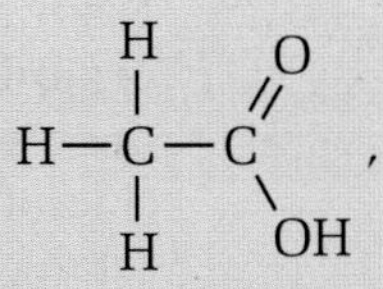

(ethanoic acid structure with H, C, H on left; O double bond and OH on right)

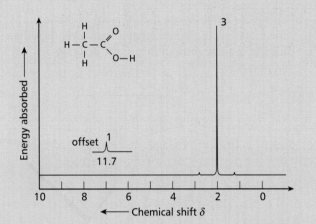

a Say which peak is caused by which hydrogen atoms in the molecule.

b How would the spectrum differ if the ethanoic acid were shaken with deuterium oxide?

2 a Give the structural formula of (i) propan-1-ol, (ii) methanal.

   b Which of the NMR spectra of (i) and (ii) above would be affected by shaking the sample with deuterium oxide?

   c Explain your answer to **b**.

3 In which of the following compounds will the proton NMR spectrum be affected by shaking the sample with deuterium oxide, D_2O: **a** butan-1-ol, **b** butan-2-ol, **c** butanal, **d** butanone, **e** butanoic acid? Explain your answers.

1 a From the information given, draw the structural formula of each organic compound. **All** of the compounds consist of molecules which have **three carbon atoms**.
  (i) A hydrocarbon that rapidly decolourises bromine.
  (ii) A compound that is oxidised to a ketone.
  (iii) An ester.
  (iv) A compound that forms a silver mirror when heated with Tollens' reagent.
  (v) An amino acid. [5]

b 1-bromobutane (drawn below) can be used in the organic synthesis of a range of organic compounds by making use of different types of reaction.

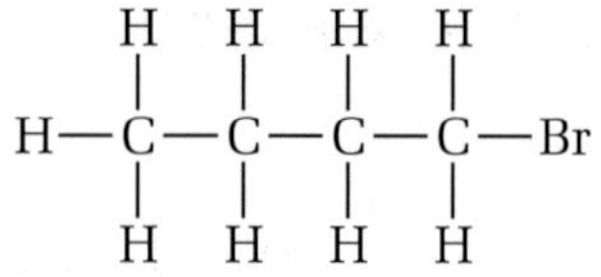

$$\begin{array}{ccccccc} & H & H & H & H & \\ & | & | & | & | & \\ H- & C- & C- & C- & C- & Br \\ & | & | & | & | & \\ & H & H & H & H & \end{array}$$

For each of the following reactions, complete a balanced equation for the reaction you have chosen. The equation should show clearly the structure of the organic product(s).

  (i) a nucleophilic substitution reaction;

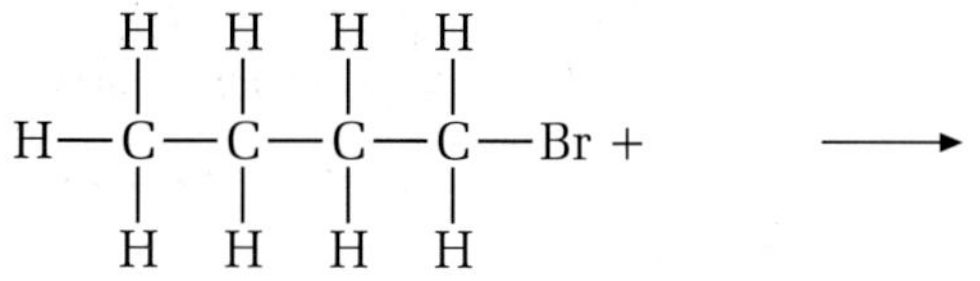

$$\begin{array}{ccccccc} & H & H & H & H & \\ & | & | & | & | & \\ H- & C- & C- & C- & C- & Br + \\ & | & | & | & | & \\ & H & H & H & H & \end{array} \longrightarrow$$

  (ii) an elimination reaction.

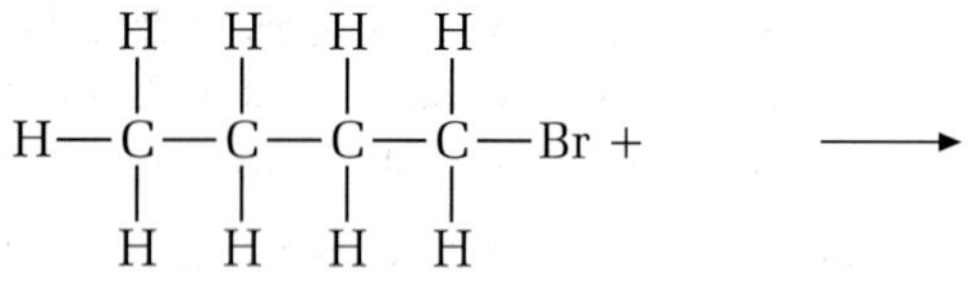

$$\begin{array}{ccccccc} & H & H & H & H & \\ & | & | & | & | & \\ H- & C- & C- & C- & C- & Br + \\ & | & | & | & | & \\ & H & H & H & H & \end{array} \longrightarrow$$

[4]

[Total: 9]

OCR, Specimen 2000

2 Benzene, methlybenzene and phenol are used in the chemical and pharmaceutical industry as starting materials for making more complex aromatic compounds.

a Methylbenzene can also be made in the laboratory from benzene and chloromethane.
  (i) Draw the structural formula of methylbenzene. [1]
  (ii) Give the equation for the preparation of methylbenzene from benzene. [1]
  (iii) Identify, by name or formula, a suitable catalyst for this reaction. [1]

  (iv) Methylbenzene is more reactive than benzene. Name and draw the structural formulae of an **organic** product which might be formed from the reaction of methylbenzene with chloromethane in the presence of the catalyst. [2]

b Complete and balance the following equations for the reactions of phenol, giving structural formulae for the organic compounds in the boxes provided.

  (i)

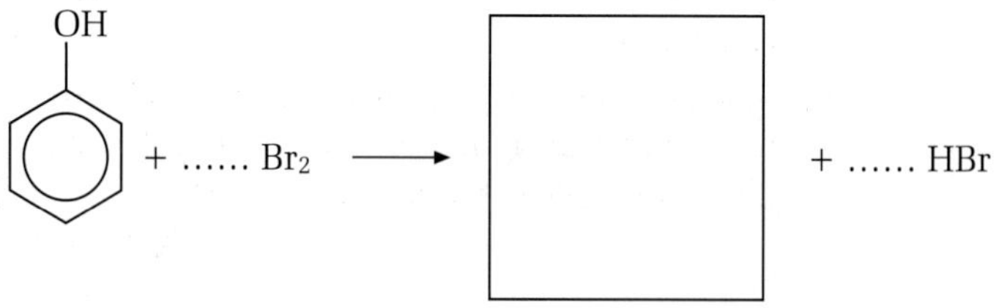

OH
⬡ + ... Br_2 ⟶ ▢ + ... HBr

[3]

  (ii)

OH
⬡ + Na ⟶ ▢ + ... H_2

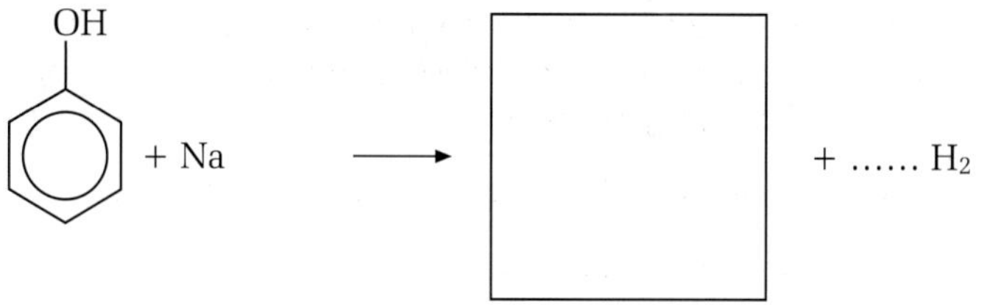

[2]

c State a general use for phenols. [1]

[Total: 11]

OCR, Jan 2003

3 Compounds **K** and **L** are structural isomers.

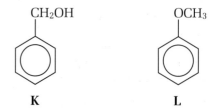

K   **L**

a (i) What is the molecular formula of these isomers? [1]
  (ii) Calculate the mass:charge ratio, m/e, you expect for the molecular ion peak in the mass spectrum of **K**, showing your working. [1]
  (iii) A sample of **L** is sent for analysis to determine its percentage by mass of carbon and hydrogen. Calculate the expected results. [2]

b Explain how infra-red spectra would allow you to distinguish between samples of **K** and **L**. [3]

c (i) Compound **K** gives the nmr spectrum below. Identify which of the protons are responsible for each peak. Explain your reasoning.

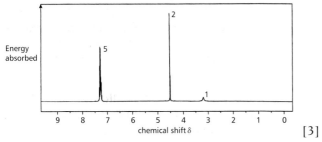

[3]

(ii) A sample of **K** is shaken with D_2O and the spectrum is re-run. Describe how the spectrum is changed. [1]

(iii) Suggest possible δ values for the peaks in the nmr spectrum of compound **L**. For each peak, give the number of protons responsible. [4]

[Total: 15]

OCR, June 2002

4 Butanone can be reduced with $NaBH_4$ to form an alcohol **G**. Compound **G** has a chiral centre and can display optical isomerism.

a (i) Explain the meaning of the term *chiral centre*.
(ii) Deduce the identify of compound **G** and draw its optical isomers. [3]

b Butanone has the infra-red spectrum below.

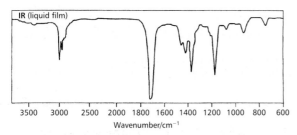

(i) How does this infra-red spectrum confirm the presence of the functional group present in butanone?
(ii) How would you expect the infra-red spectrum of compound **G** to differ from that of butanone? Explain your answer clearly? [4]

c Butanone reacts with hydrogen cyanide in the presence of potassium cyanide.
(i) Describe, with the aid of curly arrows, the mechanism for this reaction.
(i) What type of reaction is this? [4]

[Total: 11]

OCR, Specimen 2000

5 Compound **E** is an aromatic hydrocarbon with the molecular formula C_8H_{10}.

a Draw structures for the **four** possible isomers of **E**. [4]

b The nmr spectrum of **E** is shown below.

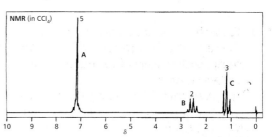

Suggest the identity of the protons responsible for the groups of peaks **A**, **B** and **C**. For each group of peaks, explain your reasoning carefully in terms of both the chemical shift value and the splitting pattern.
(i) **A** (ii) **B** (iii) **C** [9]

c Using the evidence from **b**, identify and show the structure of hydrocarbon **E**. [1]

[Total: 14]

OCR, Specimen 2000

6 a A section of a polymer has the structure shown below.

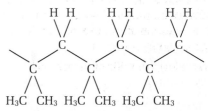

(i) Circle a repeat unit of this polymer on the diagram above. [1]
(ii) Deduce the empirical formula of this polymer. [1]
(iii) Draw a structure for a monomer from which this polymer could be made. Your structure should show any multiple bonds. [1]

b Proteins are natural polymers made from α-amino acids, such as glycine, H_2NCH_2COOH.
(i) Name the functional group made during amino acid polymerisation and draw its displayed formula. [2]
(ii) Name this type of polymerisation reaction. [1]
(iii) Draw a displayed and a skeletal formula for the dipeptide **H**, $C_4H_8N_2O_3$, made from glycine, H_2NCH_2COOH. [2]
(iv) A student made 1.10 g of dipeptide **H** starting from 1.40 g of glycine.
Calculate the percentage yield obtained. Give your answer to 3 significant figures. [4]
(v) When glycine is treated with hydrochloric acid a compound **J**, $C_2H_6ClNO_2$, is formed. Draw a structure for compound **J**. [2]

[Total: 14]

OCR, June 2002

Trends and patterns

8 Lattice enthalpy

8.1 Hess's Law

In *Essential AS Chemistry for OCR*, Chapter 13, we used Hess's Law to construct **enthalpy** cycles and enthalpy diagrams. In this module we return to Hess's Law and use it to investigate the enthalpy changes when an ionic compound is formed. The enthalpy changes you will use are defined in the box *Definition of terms*.

8.1.1 Ionic bonding

In a simple model of ionic bonding, electrons are transferred from metal atoms to non-metal atoms. Positively charged metal ions and negatively charged non-metal ions are formed that all have a full outer shell of electrons. These ions arrange themselves into a lattice so that ions of opposite charge are next to one another.

DEFINITION OF TERMS

Standard conditions are 298 K and 100 kPa.

1. **Standard enthalpy change of formation, $\Delta H_f^{\ominus}$** is the enthalpy change for the formation of 1 mole of a compound from its elements under standard conditions.

For example,

$$H_2(g) + \tfrac{1}{2}O_2(g) \rightarrow H_2O(l) \quad \Delta H_f^{\ominus} = -286 \text{ kJ mol}^{-1}$$

2. **Standard enthalpy change of atomisation, $\Delta H_{at}^{\ominus}$** is the enthalpy change for the formation of one mole of gaseous atoms from the element in its standard state under standard conditions.

For example,

$$\tfrac{1}{2}Cl_2(g) \rightarrow Cl(g) \quad \Delta H_{at}^{\ominus} = +121 \text{ kJ mol}^{-1}$$

Note that this is given per mole of chlorine *atoms* and not per mole of chlorine molecules. It is particularly important to have an equation to refer to, for this type of enthalpy change.

3. **Standard enthalpy change of first ionisation (first ionisation energy, 1st IE)** is the enthalpy change when one mole of gaseous atoms is converted into a mole of gaseous ions each with a single positive charge under standard conditions.

For example,

$$Na(g) \rightarrow Na^+(g) + e^- \quad \Delta H^{\ominus} = +496 \text{ kJ mol}^{-1}$$

Note that the enthalpy change of second ionisation (second ionisation energy, 2nd IE) refers to the loss of a mole of electrons from a mole of singly positively charged ions under standard conditions.

For example,

$$Na^+(g) \rightarrow Na^{2+}(g) + e^- \quad \Delta H^{\ominus} = +4563 \text{ kJ mol}^{-1}$$

4. **Standard enthalpy change of electron gain (first electron affinity)** is the enthalpy change when a mole of gaseous atoms is converted to a mole of gaseous ions each with a single negative charge under standard conditions.

For example,

$$O(g) + e^- \rightarrow O^-(g) \quad \Delta H^{\ominus} = -141.1 \text{ kJ mol}^{-1}$$

Note that this refers to single atoms, not to oxygen molecules O_2.

The second electron affinity is the enthalpy change when a mole of electrons is added to a mole of gaseous ions each with a single negative charge under standard conditions.

For example,

$$O^-(g) + e^- \rightarrow O^{2-}(g) \quad \Delta H^{\ominus} = +798 \text{ kJ mol}^{-1}$$

5. **Standard enthalpy change of lattice formation**, often called lattice energy or LE, is the enthalpy change when one mole of ionic compound is formed from its gaseous ions under standard conditions.

For example,

$$Na^+(g) + Cl^-(g) \rightarrow NaCl(s) \quad LE = -788 \text{ kJ mol}^{-1}$$

The removal of electrons from metals requires energy to be put in. (It is an **endothermic** process). But most ionic compounds form readily and give out energy as they form (an **exothermic** process). To see why this happens we need to look at *all* the energy changes that occur when ionic compounds are formed.

8.1.2 Enthalpy changes on forming ionic compounds

If a cleaned piece of solid sodium is placed in a gas jar containing chlorine gas, a rapid exothermic reaction takes place, forming solid sodium chloride.

$$Na(s) + \tfrac{1}{2}Cl_2(g) \rightarrow (Na^+ + Cl^-)(s) \quad \Delta H_f^{\ominus} = -411 \text{ kJ mol}^{-1}$$

This is the overall reaction. We can think of it as taking place in several steps:

- At some stage the sodium atom must give up an electron to form Na^+.

$$Na(g) \rightarrow Na^+(g) + e^-$$

The enthalpy change for this process is the enthalpy change of first ionisation (ionisation energy, 1st IE) of sodium and is $+496 \text{ kJ mol}^{-1}$, i.e. energy must be *put in* for this process to occur.

- At some stage a chlorine atom must gain an electron:

$$Cl(g) + e^- \rightarrow Cl^-(g)$$

The enthalpy change for this process of electron *gain* is called the first electron affinity, 1st EA. The first electron affinity for the chlorine atom is -349 kJ mol^{-1}, i.e. energy is given out when this process occurs.

But there is more. The reaction we are considering involves *solid* sodium, not gaseous, and chlorine *molecules*, not separate atoms, so we must include the energy changes for the processes below:

$$Na(s) \rightarrow Na(g) \qquad \Delta H_{at}^{\ominus} = +108 \text{ kJ mol}^{-1}$$
$$\tfrac{1}{2}Cl_2(g) \rightarrow Cl(g) \qquad \Delta H_{at}^{\ominus} = +122 \text{ kJ mol}^{-1}$$

These energy values are called atomisation enthalpies, $\Delta H_{at}^{\ominus}$. Notice that energy has to be *put in* to 'pull apart' the atoms ($\Delta H_{at}^{\ominus}$ is positive in both cases).

There is a further energy change to be added. At room temperature sodium chloride exists as a solid lattice of alternating positive and negative ions and not as separate gaseous ions. If positively charged ions come together with negatively charged ions, they form a solid lattice and energy is given out owing to the attraction of the ions. This is called the lattice enthalpy, LE, and it refers to the process:

$$Na^+(g) + Cl^-(g) \rightarrow (Na^+ + Cl^-)(s) \qquad LE = -788 \text{ kJ mol}^{-1}$$

Lattice enthalpy is in effect a measure of the strength of an ionic bond. The bigger the numerical value of the lattice enthalpy (i.e. the more negative it is), the stronger the bond.

So we have five processes that lead to the formation of NaCl(s) from its elements.

$$Na(g) + \tfrac{1}{2}Cl_2(g) \rightarrow (Na^+ + Cl^-)(s) \qquad \Delta H_f^{\ominus} = -411 \text{ kJ mol}^{-1}$$

These are:

- Atomisation of Na

$$Na(s) \rightarrow Na(g) \qquad \Delta H_{at}^{\ominus} = +108 \text{ kJ mol}^{-1}$$

> **HINT**
>
> Do not confuse ionisation energy that refers to electron *loss* with electron affinity, which refers to electron *gain*.

69

- Atomisation of Cl_2

$$\tfrac{1}{2} Cl(g) \rightarrow Cl(g) \qquad \Delta H^{\ominus}_{at} = +122 \text{ kJ mol}^{-1}$$

- Ionisation (e^- loss) of Na

$$Na(g) \rightarrow Na^+(g) + e^- \qquad \text{first IE} = +496 \text{ kJ mol}^{-1}$$

- Gain of e^- by Cl

$$Cl(g) + e^- \rightarrow Cl^-(g) \qquad \text{first EA} = -349 \text{ kJ mol}^{-1}$$

- Formation of lattice

$$Na^+(g) + Cl^-(g) \rightarrow (Na^+ + Cl^-)(s) \quad LE = -788 \text{ kJ mol}^{-1}$$

Hess's law tells us that the total energy (or enthalpy) change for a chemical reaction is the same *whatever route is taken*, provided that the initial and final conditions are the same. So the sum of the first five energy changes is equal to the enthalpy change of formation of sodium chloride.

We can calculate any of the quantities, provided all the others are known. We do this by using a thermochemical cycle called a Born–Haber cycle, (the same Haber as in the Haber process), see Topic 8.2.

QUICK QUESTIONS

1 **a** Draw a dot-and-cross diagram to show the electron transfer when sodium chloride is formed from its elements.

b Which two of the enthalpy changes involved in the formation of sodium chloride from its elements are represented in the dot-and-cross diagram?

c Which of the enthalpy changes involved in the formation of sodium chloride were not represented?

2 Explain why

a loss of an electron from a sodium atom (ionisation) is an endothermic process (energy has to be put in).

b gain of an electron by a chlorine atom is an exothermic process (energy is given out).

3 **a** Magnesium normally forms Mg^{2+} ions. Write the equation to represent
(i) the first
(ii) the second ionisation of magnesium.

b In terms of the energies for the two processes above, what is the energy change when a Mg^{2+} ion is formed from a Mg atom?

Born–Haber cycles

A Born–Haber cycle is a thermochemical cycle that includes all the enthalpy changes involved in the formation of an ionic compound. We construct it by starting with the elements in their standard states. All elements in their standard states have zero enthalpy by definition.

8.2.1 The Born–Haber cycle for sodium chloride

There are a total of six steps in the Born–Haber cycle for sodium chloride, see Topic 8.1. Here we will use the cycle to calculate the lattice enthalpy. The other five steps are shown in the margin. (Remember that if we know any five, we can calculate the other). Figure 8.1 shows you how each step is added to the one before, starting from the elements in their standard state. Positive (endothermic changes) go upward and negative exothermic changes) go downward.

$$Na(s) \rightarrow Na(g)$$
$$\Delta H_{at}^{\ominus} = +108 \text{ kJ mol}^{-1}$$
$$\tfrac{1}{2} Cl_2(g) \rightarrow Cl(g)$$
$$\Delta H_{at}^{\ominus} = +122 \text{ kJ mol}^{-1}$$
$$Na(g) \rightarrow Na^+(g) + e^-$$
$$\text{first IE} = +496 \text{ kJ mol}^{-1}$$
$$Cl(g) + e^- \rightarrow Cl^-(g)$$
$$\text{first EA} = -349 \text{ kJ mol}^{-1}$$
$$Na(s) + \tfrac{1}{2} Cl_2(g) \rightarrow (Na^+ + Cl^-)(s)$$
$$\Delta H_f^{\ominus} = -411 \text{ kJ mol}^{-1}$$

When drawing Born–Haber cycles:

- Make up a rough scale, e.g. 1 line of lined paper to 100 kJ mol^{-1}. Plan out roughly first to avoid going off the top or bottom of the paper. (The zero line representing elements in their standard state will need to be in the middle of the paper.)

- It is better not to put $+$ and $-$ signs in on the diagram, because if you do you can easily count them twice. If you go from a lower line to a higher line, the enthalpy change is positive. If you go from a higher line to a lower, the enthalpy change is negative. So, the quantities can be given signs after you have calculated their sizes. For example, the lattice enthalpy refers to the change *from* separate gaseous ions *to* crystalline solid (downhill) so it has a negative sign.

Using a Born–Haber cycle we can see why the formation of an ionic compound from its elements is an exothermic process – the large amount of energy given out when the lattice is formed is greater than the energy that has to be put in to form the positive ions.

8.2.2 The Born–Haber cycle for magnesium chloride

Figure 8.2 is the complete Born–Haber cycle for magnesium chloride, MgCl$_2$, together with notes on how it is constructed.

Since *two* chlorines are involved all the quantities related to Cl are doubled.

i.e. $2 \times \Delta H_{at}^{\ominus}$, $2 \times$ first electron affinity (*not* 1st + 2nd electron affinities)

HINT

First electron affinities of *all* elements are negative because the added electron is attracted by the nuclear charge. Second (and subsequent) electron affinities are *always* positive because a negatively charged electron is being added to an already negatively charged ion.

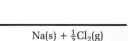
$$Na(s) + \tfrac{1}{2}Cl_2(g)$$

1. Elements in their standard states. This is the energy zero of the diagram

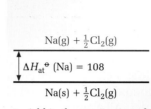
$$Na(g) + \tfrac{1}{2}Cl_2(g)$$
$$\Delta H_{at}^{\ominus}(Na) = 108$$
$$Na(s) + \tfrac{1}{2}Cl_2(g)$$

2. Add in the atomisation of sodium. This is positive, so it is drawn 'uphill'

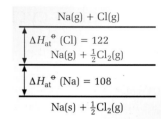
$$Na(g) + Cl(g)$$
$$\Delta H_{at}^{\ominus}(Cl) = 122$$
$$Na(g) + \tfrac{1}{2}Cl_2(g)$$
$$\Delta H_{at}^{\ominus}(Na) = 108$$
$$Na(s) + \tfrac{1}{2}Cl_2(g)$$

3. Add in the atomisation of chlorine. This too is positive and so drawn 'uphill'

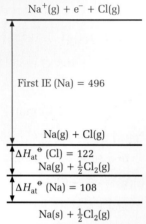

4. Add in the ionisation of sodium, also positive and so drawn 'uphill'

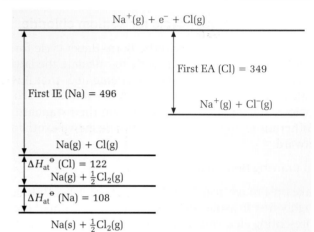

5. Add in the electron affinity of chlorine. This is a negative energy change and so is drawn 'downhill'

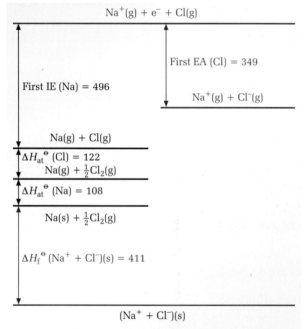

6. Add in the enthalpy of formation of sodium chloride, also negative and drawn 'downhill'

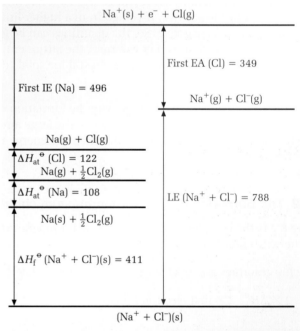

7. The final unknown quantity is the lattice energy of sodium chloride. The size of this is 788 kJ mol^{-1} from the diagram. The definition of lattice energy is the change from separate ions to solid lattice and we must therefore go 'downhill', so LE (Na$^+$ + Cl$^-$)(s) = –788 kJ mol^{-1}

The complete Born–Haber cycle for sodium chloride, NaCl

Fig 8.1 *Stages in the construction of the Born–Haber cycle to find the lattice enthalpy of sodium chloride, NaCl. All energies are in kJ mol^{-1}*

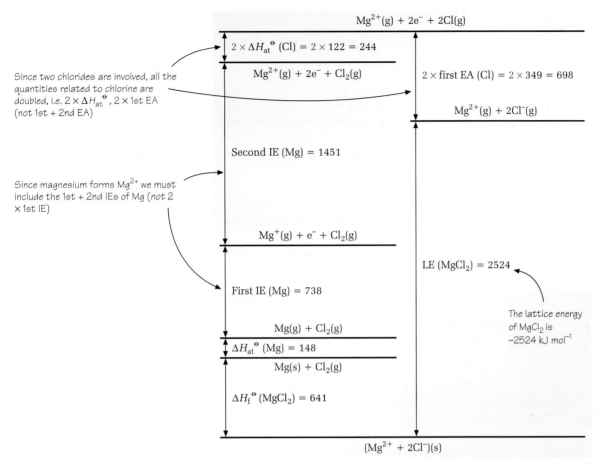

Fig. 8.2 *The Born–Haber cycle for magnesium chloride, MgCl$_2$. All energies in kJ mol^{-1}*

QUICK QUESTIONS

1 a Draw a Born–Haber cycle to find the lattice enthalpy for sodium fluoride, NaF, The values for the relevant enthalpy terms are given below.

b What is the lattice enthalpy of NaF?

$Na(s) \rightarrow Na(g)$
$\Delta H_{at}^{\ominus} = +108 \text{ kJ mol}^{-1}$

$\frac{1}{2} F_2(g) \rightarrow F(g)$
$\Delta H_{at}^{\ominus} = +79 \text{ kJ mol}^{-1}$

$Na(g) \rightarrow Na^+(g) + e^-$
first IE $= +496 \text{ kJ mol}^{-1}$

$F(g) + e^- \rightarrow F^-(g)$
first EA $= -328 \text{ kJ mol}^{-1}$

$Na(s) + \frac{1}{2} F_2(g) \rightarrow (Na^+ + F^-)(s)$
$\Delta H_f^{\ominus} = -574 \text{ kJ mol}^{-1}$

Trends

There are patterns in the sizes of the lattice enthalpies of compounds which are related to the size and charge of the ions involved. We can see this pattern by looking at the lattice enthalpies of the Group 1 and Group 2 halides.

The size of the metal ion also has an effect on the temperature at which the metal carbonates in Group 2 decompose.

8.3.1 The lattice enthalpies of halides

Tables 8.1 and 8.2 show the lattice enthalpies for the halides of the metals in Groups 1 and 2 of the Periodic Table.

Table 8.1 Lattice enthalpies for M^+X^- in kJ mol^{-1}

		Larger negative ions (anions) →			
		F$^-$	**Cl$^-$**	**Br$^-$**	**I$^-$**
Larger positive ions (cations)	Li$^+$	−1031	−848	−803	−759
	Na$^+$	−918	−780	−742	−705
	K$^+$	−817	−711	−679	−651
	Rb$^+$	−783	−685	−656	−628
	Cs$^+$	−747	−661	−635	−613

Table 8.2 Lattice enthalpies for $M^{2+}X^{2-}$ in kJ mol^{-1}

		Larger anions →	
		O^{2-}	**S^{2-}**
Larger cations	Be^{2+}	−4443	−3832
	Mg^{2+}	−3791	−3299
	Ca^{2+}	−3401	−3013
	Sr^{2+}	−3223	−2843
	Ba^{2+}	−3054	−2725

There are two patterns:

- The larger the ions, the smaller the numerical value of the lattice enthalpy, which means that less energy is given out when the lattice forms. (Take care here; as lattice enthalpies are all negative, we must be careful to say *the numerical value*. If we say the lattice enthalpies are smaller, we could mean more negative.)

- More highly charged ions give rise to lattice enthalpies with *much* greater numerical values, which means that more energy is given out when the lattice forms. For example compare Li$^+$ F$^-$ and Be^{2+}O^{2-} (compounds with ions of comparable sizes). Doubling the charges increases the numerical value of the lattice enthalpy more than fourfold.

We can explain these patterns as follows.

Size

Larger ions do not pack together as closely as small ions, and so they attract each other less strongly than the smaller ions. This means we need to put in less energy to break up the lattice (and also get less energy out when the lattice forms).

Charge

More highly charged ions attract each other more strongly. This means we must put in more energy to break up the lattice (and thus get more energy out when it forms). In fact the laws of electrostatic attraction say that the attraction is proportional to the size of both the negative and the positive charge which is why doubling both charges leads to an approximately fourfold increase in the numerical value of the lattice energy.

HINT

In general, a compound with a high numerical value of lattice energy will be difficult to melt (because to melt a compound, its ions must be pulled apart). In other words, the compound is heat resistant. This is part of the reason why magnesium oxide (melting point 2853°C) is used to line the inside of furnaces. Heat resistant compounds are described as refractory.

8.3.2 Thermal decomposition of Group 2 metal carbonates

The carbonates of Group 2 metal have the formula MCO_3 or, ionically, $M^{2+} CO_3^{2-}$, where M represents any metal. When they are heated the carbonates break down ('**thermally decompose**') to give the metal oxide and carbon dioxide:

$$MCO_3(s) \rightarrow MO(s) + CO_2(g)$$

The temperature at which this occurs is shown in Table 8.3.

There is a clear trend; it becomes harder to decompose the carbonates as we go down the group and the metal ion gets bigger.

The ionic equation is the same for all the carbonates. We can see that nothing happens to the metal ion in the reaction – it is the carbonate ion that changes, decomposing into an oxide ion and carbon dioxide:

$$(M^{2+} CO_3^{2-})(s) \rightarrow (M^{2+} + O^{2-})(s) + CO_2(g)$$

However, the metal ion does influence the reaction by **polarising** the carbonate ion (see *Essential AS Chemistry for OCR*, Topic 3.4.3). Smaller metal ions can get closer to the carbonate ions and polarise them more, so in magnesium carbonate, the carbonate ions are significantly distorted:

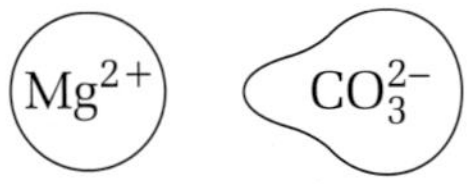

Table 8.3 *Decomposition temperatures of the carbonates of Group 2 metals*

Compound	Decomposition temperature/°C
$MgCO_3$	350
$CaCO_3$	830
$SrCO_3$	1340
$BaCO_3$	1450

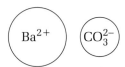

The more distorted the shape, the easier it is to decompose the carbonate ion. Larger ions, such as barium, hardly distort the carbonate ion at all:

This explains why magnesium carbonate decomposes at a much lower temperature than barium carbonate.

If we look at the movement of electrons we see that electrons have to be dragged towards the oxygen atom that is to become the oxide ion. A small positive ion close to the carbonate ion helps this to happen:

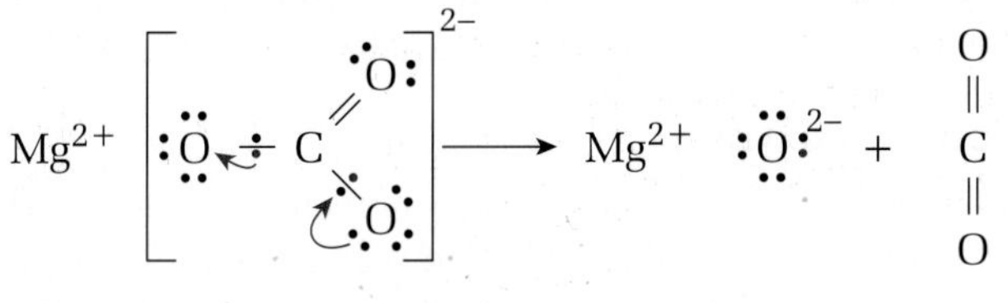

QUICK QUESTIONS

1. The lattice enthalpy of lithium fluoride, LiF, is -1031 kJ mol^{-1}. The lattice energy of sodium fluoride, NaF, is -918 kJ mol^{-1}. Explain why the lattice enthalpy of LiF is numerically larger than that of NaF.

2. The lattice enthalpy of lithium chloride, LiCl, is -845 kJ mol^{-1}. Explain why this is numerically smaller than that for LiF.

3. The lattice enthalpy of magnesium oxide, MgO, is -3791 kJ mol^{-1}. The lattice enthalpy of sodium fluoride, NaF, is -918 kJ mol^{-1}. Explain why the lattice enthalpy of MgO is approximately four times numerically the lattice enthalpy of NaF.

4. Why is comparing the lattice enthalpy of MgO with NaF more meaningful than comparing the lattice enthalpy of MgO with that of KCl?

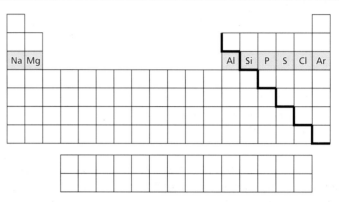

As we move across a **period** in the Periodic Table from left to right, there are a number of trends in the properties of the elements. In this chapter we shall look at Period 3, the elements from sodium to argon, and also see the trends in the compounds of these elements.

9.1.1 The elements

The most obvious trend in the elements is from metals on the left to non-metals on the right.

- Sodium, magnesium and aluminium are metallic – they are shiny (when freshly exposed to air) conduct electricity and react with dilute acids to give hydrogen and salts.
- Silicon is a semi-metal (or metalloid) – it conducts electricity to some extent, a property that is useful in making semiconductor devices.
- Phosphorus, sulphur and chlorine are typical non-metals – in particular, they do not conduct electricity.
- Argon is a noble gas, being chemically unreactive and existing as separate atoms.

9.1.2 The redox reactions of the elements

The reactions of the elements in Period 3 are all redox reactions since every element starts with an oxidation number of zero, and, after it has reacted, ends up with a positive or a negative oxidation number.

9.1.3 Reaction with oxygen

All the elements in Period 3 (except for argon) are relatively reactive. Their oxides can all be prepared by direct reaction of the element with oxygen. As we go across the period, the oxides show trends, for example, strongly basic on the left of the Periodic .Table, acidic on the right, see Table 9.1. Topic 9.3 looks at this in more detail.

Magnesium

A strip of magnesium ribbon burning in air, burns with an even more intense white flame if it is lowered into a gas jar of oxygen. The white powder that is produced is magnesium oxide, see the photograph opposite.

$$\text{magnesium} + \text{oxygen} \rightarrow \text{magnesium oxide}$$

$$\overset{0}{2\text{Mg(s)}} + \overset{0}{\text{O}_2\text{(g)}} \rightarrow \overset{+\text{II} -\text{II}}{2\text{MgO(s)}}$$

The oxidation numbers show how magnesium has been oxidised (its oxidation number has increased) and oxygen has been reduced (its oxidation number has decreased).

Magnesium burning in oxygen

Table 9.1 *Trends across Period 3*

Group	1	2	3	4	5	6	7	0
Element	Na	Mg	Al	Si	P	S	Cl	Ar
	Metals			Semi-metal	Non-metals			(Noble gas)
	less reactive →				← less reactive			
Structure of element	Giant metallic			Giant covalent	Molecular			(Atomic)
Ion formed	Na⁺	Mg²⁺	Al³⁺	None	P³⁻	S²⁻	Cl⁻	none

Magnesium oxide is an ionic compound and dissolves slightly in water to form a somewhat alkaline solution of pH about 10. It is a basic oxide.

Aluminium

When aluminium powder is heated and then lowered into a gas jar of oxygen, it burns brightly to give aluminium oxide – a white powder, see photograph below.

$$\text{aluminium} + \text{oxygen} \rightarrow \text{aluminium oxide}$$
$$\underset{0}{} \quad \underset{0}{} \quad \underset{+\text{III} \ -\text{II}}{}$$
$$4\text{Al}(s) + 3\text{O}_2(g) \rightarrow 2\text{Al}_2\text{O}_3(s)$$

Aluminium burning in oxygen: powdered aluminium is being sprinkled into the flame

> **NOTE**
>
> Note that the sum of the oxidation numbers in Al_2O_3 is zero (as it is in all compounds) $(2 \times \text{III}) + (3 \times -\text{II}) = 0$

Aluminium is a reactive metal but it is always coated with a surface layer of oxide, which protects it from further reaction. This is why aluminium appears to be an unreactive metal and is used for many everyday purposes.

Aluminium oxide is principally ionic but has some degree of covalent bonding and will not dissolve in water. It is an amphoteric oxide, which means that it shows both acidic and basic properties.

Sulphur

When sulphur powder is heated and lowered into a gas jar of oxygen, it burns with a blue flame to form the colourless gas sulphur dioxide (and a little sulphur trioxide also forms), see photograph. Sulphur dioxide dissolves in water to form an acidic solution.

$$\text{sulphur} + \text{oxygen} \rightarrow \text{sulphur dioxide}$$
$$\underset{0}{} \quad \underset{0}{} \quad \underset{+\text{IV} \ -\text{II}}{}$$
$$\text{S}(s) + \text{O}_2(g) \rightarrow \text{SO}_2(g)$$

In all these redox reactions the oxidation number of the Period 3 element increases and that of the oxygen decreases (from 0 to –II in each case). The oxidation number changes are shown in red in the symbol equations above. Notice how the oxidation number of the Period 3 element in the oxide increases as we move from left to right across the period.

Sulphur burning in oxygen

> ### QUICK QUESTIONS
>
> 1 Metals are shiny, conduct electricity and if they are reactive, react with acids to give hydrogen. Give three more properties not mentioned in the text.
>
> 2 Non-metals do not conduct electricity. Give two more properties typical of non-metals.
>
> 3 Magnesium oxide is an ionic compound. Give the charges of the two ions of magnesium oxide.
>
> 4 a Draw a diagram to show the electron transfer in the bonding in magnesium oxide.
>
>   b Give one property of magnesium oxide that is typical of an ionically bonded compound.
>
>   c Give one property of magnesium oxide that is typical of a compound with a giant structure.
>
> 5 Sulphur dioxide has a covalently bonded molecular structure.
>
>   a Give one property of sulphur dioxide that is typical of a covalently bonded compound.
>
>   b Give one property of sulphur dioxide that is typical of a compound with a molecular structure.

77

More reactions of Period 3 elements

9.2.1 Reaction with chlorine

All the elements in Groups 1 to 6 in Period 3 will react directly with chlorine to form chlorides. In each case a little of the element is heated in air and lowered into a gas jar of chlorine where the reaction takes place.

Alternatively, the element can be heated in a stream of chlorine and the chloride collected as it cools, see Figures 9.1 and 9.2.

The method of collection differs slightly because:

- the metal chlorides have high melting points and need no cooling to be collected as solids;
- the non-metal chlorides have low melting points and need cooling in order to be collected – silicon tetrachloride is a liquid at room temperature and phosphorus pentachloride is a solid that sublimes (turns straight from liquid to gas) at 162°C.

The difference in melting points is because the chlorides of sodium and magnesium are ionic compounds and therefore have giant structures, whereas the chlorides of silicon and phosphorus are simple molecular.

The equations are:

$$\overset{0}{2Na(s)} + \overset{0}{Cl_2(g)} \rightarrow \overset{+I\ -I}{2NaCl(s)}$$

$$\overset{0}{Mg(s)} + \overset{0}{Cl_2(g)} \rightarrow \overset{+II\ -I}{MgCl_2\,(s)}$$

$$\overset{0}{2Al(s)} + \overset{0}{3Cl_2(g)} \rightarrow \overset{+III\ -I}{2AlCl_3(s)}$$

$$\overset{0}{Si(s)} + \overset{0}{2Cl_2(g)} \rightarrow \overset{+IV\ -I}{SiCl_4(l)}$$

$$\overset{0}{2P(s)} + \overset{0}{5Cl_2(g)} \rightarrow \overset{+V\ -I}{2PCl_5(s)}$$

Notice that the oxidation numbers are given per atom, so that, for example in $SiCl_4$ the sum of the oxidation numbers is zero $(1 \times +IV) + (4 \times -I) = 0$.

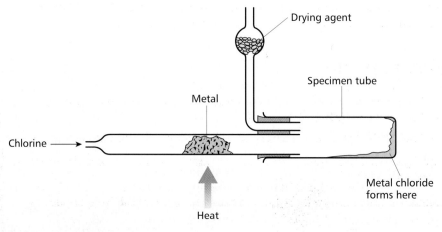

Fig 9.1 *Apparatus for preparing metal chlorides*

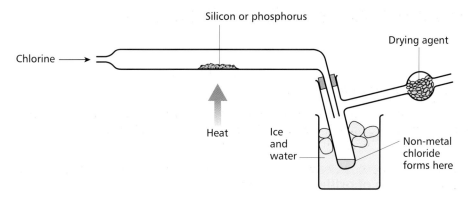

Fig 9.2 *Apparatus for preparing non-metal chlorides*

These are all redox reactions in which the oxidation number of the chlorine drops from 0 to –I and that of the other element goes up. The oxidation number changes are shown in red in the symbol equations above. Notice how the oxidation number of the Period 3 element in the chloride increases as we move from left to right across the period.

9.2.2 Reactions with water

Both sodium and magnesium react with water to form the hydroxide and hydrogen.

Sodium

The reaction of sodium is vigorous – the sodium floats on the surface of the water and fizzes rapidly, melting because of the heat given out by the reaction. A strongly alkaline solution of sodium hydroxide is formed (pH 12–14).

$$\overset{0}{2Na(s)} + \overset{+I\ -II}{2H_2O(l)} \rightarrow \overset{+I\ -II\ +I}{2NaOH(aq)} + \overset{0}{H_2(g)}$$

Magnesium

The reaction of magnesium is very slow at room temperature, only a few bubbles of hydrogen are formed after some days. The resulting solution is less alkaline than in the case of sodium because magnesium hydroxide is only sparingly soluble (pH around 10).

$$\overset{0}{Mg(s)} + \overset{+I\ -II}{2H_2O(l)} \rightarrow \overset{+II\ -II\ +I}{Mg(OH)_2(aq)} + \overset{0}{H_2(g)}$$

The reactions are redox ones in which the oxidation number of the metal increases and that of some of the hydrogen atoms decreases. The oxidation number changes are shown in red in the symbol equations above.

QUICK QUESTIONS

1   **a**   What pattern can you see in the formulae of the chlorides as we cross Period 3 from left to right?

   b   Predict the formula of the chloride of sulphur if this pattern continues.

2   Explain why the elements in Groups 7 and 0 in Period 3 do not react directly with chlorine.

3   Show that the sum of the oxidation numbers in magnesium hydroxide is zero.

The compounds of elements in Period 3

There are some important trends in the reactions of the Period 3 compounds. These trends are a result of the change from metals on the left of the Periodic Table to non-metals on the right. The reactions of the oxides and chlorides with water show the trends.

9.3.1 Reaction of oxides with water

- Metals on the left of the Periodic Table typically have basic oxides. For example:

  Magnesium oxide reacts with water to give magnesium hydroxide, which is sparingly soluble in water and produces a somewhat alkaline solution of pH about 10:

  $$MgO(s) + H_2O(l) \rightarrow Mg(OH)_2(s) \rightleftharpoons Mg^{2+}(aq) + 2OH^-(aq)$$

- As we move to the right along the period we reach aluminium:

  Aluminium oxide is virtually insoluble in water. But, aluminium oxide will react with both acids and alkalis, so it is an **amphoteric** oxide.

- Non-metals on the right of the Table typically have acidic oxides. For example:

  Sulphur dioxide reacts with water to give a strongly acidic solution of sulphuric(IV) acid (sulphurous acid). This dissociates producing H^+ ions, which cause the acidity of the solution:

  $$SO_2(g) + H_2O(l) \rightarrow H_2SO_3(aq) \rightarrow H^+(aq) + HSO_3^-(aq)$$

The overall pattern is that metal oxides (on the left of the period) form alkaline solutions in water and non-metal oxides (on the right of the period) form acidic ones. Semi-metals (in the middle of the period) show amphoteric behaviour. You can see this pattern in Table 9.2 which gives all the elements in Period 3 for comparison.

Table 9.2 Trends in the properties of the oxides of Period 3 elements

Group	1	2	3	4	5	6	7	0
Element	Na	Mg	Al	Si	P	S	Cl	Ar
		Metals		Semi-metal		Non-metals		(Noble gas)
Oxide	Na_2O	MgO	Al_2O_3	SiO_2	P_4O_{10} (P_4O_6)	SO_2 (SO_3)	Cl_2O (Cl_2O_7 etc.)	none
		Strongly basic		Amphoteric		Acidic		
Structure of oxide		Giant ionic		Giant covalent		Molecular		None

9.3.2 Reaction of the chlorides with water

The pattern is that chlorides of metals (on the left of the period) dissolve to form neutral solutions of their ions while chlorides of non-metals (on the right of the period) react with water (hydrolyse) to form acidic solutions.

Sodium chloride and magnesium chloride

Neither sodium chloride nor magnesium chloride reacts with water. They simply dissolve to form aqueous solutions containing the metal ions and chloride ions:

$$NaCl(s) + aq \rightarrow Na^+(aq) + Cl^-(aq)$$
$$MgCl_2(s) + aq \rightarrow Mg^{2+}(aq) + 2Cl^-(aq)$$

The resulting solutions are neutral. All that has happened is that the giant ionic structures of sodium chloride and magnesium chloride have broken up to form aqueous ions.

Aluminium chloride

Aluminium chloride reacts with water to form almost-insoluble aluminium hydroxide and a solution containing H^+ ions and Cl^- ions. The H^+ ions make the solution acidic – in effect it is a solution of hydrochloric acid.

$$AlCl_3(s) + 3H_2O(l) \rightarrow Al(OH)_3(s) + 3H^+(aq) + 3Cl^-(aq)$$

Reaction with water is called hydrolysis.

This formation of an acid solution is more typical of a non-metal than a metal and reflects aluminium's position close to the 'staircase line' that separates metals from non-metals in the Periodic Table.

Silicon tetrachloride

Silicon tetrachloride also reacts with water to form an acidic solution. In this case the other product is insoluble silicon dioxide (the main constituent of sand). The reaction is vigorous, and fumes of hydrogen chloride are given off:

$$SiCl_4(s) + 2H_2O(l) \rightarrow SiO_2(s) + 4HCl(aq)$$

Some of the hydrogen chloride is given off as a gas, the rest remains in the solution and dissociates to give a solution of hydrochloric acid.

$$HCl(aq) \rightarrow H^+(aq) + Cl^-(aq)$$

Phosphorus pentachloride

Phosphorus pentachloride also reacts with water to form an acidic solution:

$$PCl_5(s) + 4H_2O(l) \rightarrow H_3PO_4(aq) + 5H^+(aq) + 5Cl^-(aq)$$

H_3PO_4 is called phosphoric(V) acid.

You can see the trend in Table 9.3, which includes all the elements in Period 3 for a comparison.

Table 9.3 The chlorides of Period 3 and their reaction with water

	Formula	NaCl	MgCl$_2$	AlCl$_3$	SiCl$_4$	PCl$_5$	SCl$_2$	ClCl
	Structure	Giant ionic	Giant ionic	Molecular	Molecular	Molecular	Molecular	Molecular
Period 3	**Effect of water**	Dissolves to gives ions Na$^+$(aq) + Cl$^-$(aq)	Dissolves to give ions Mg^{2+}(aq) + 2Cl$^-$(aq)	Hydrolyses to give ions Al(OH)$_3$(s) + 3H$^+$(aq) + 3Cl$^-$(aq)	Hydrolyses to give SiO$_2$(s) + 4H$^+$(aq) + 4Cl$^-$(aq)	Hydrolyses to give H$_3$PO$_4$(aq) + 5H$^+$(aq) + 5Cl$^-$(aq)	Hydrolyses to give S(s) + H$^+$(aq) + Cl$^-$(aq)	Partially hydrolyses to give HClO(aq) + H$^+$(aq) + Cl$^-$(aq)

Neutral (NaCl, MgCl$_2$)   Acidic (AlCl$_3$ to ClCl)

QUICK QUESTIONS

1 Sodium oxide, Na$_2$O, reacts with water in a similar way to magnesium oxide except that the product, sodium hydroxide, is more soluble than magnesium hydroxide.

   a Write a balanced equation for the reaction of sodium oxide with water.

   b (i) State the oxidation number of sodium before and after the reaction.
      (ii) Has the sodium been oxidised, reduced or neither?

2 What is the trend in solubilities of the hydroxides of the metals (i.e. Na to Al) as we cross Period 3 from left to right?

3 Suggest how we could show that the solutions formed when sodium chloride and magnesium chloride dissolve in water contain ions.

4 a What ion is responsible for the alkalinity of solutions of sodium hydroxide and magnesium hydroxide?

   b What range of pH values represents an alkaline solution?

5 a What ion is responsible for the acidity of solutions of aluminium chloride, silicon tetrachloride and phosphorus pentachloride?

   b What range of pH values represents an acidic solution?

Periodic Table: Transition elements

10.1 d-block elements

Some transition metals in use

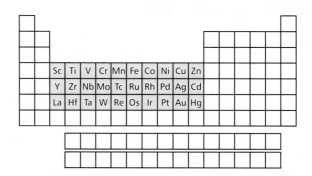

The **d-block elements** are typical metals. They are good conductors of heat and electricity. They are hard, strong and shiny, and have high melting and boiling temperatures. One notable exception is mercury, which is liquid at room temperature. There is no simple explanation for this.

These physical properties, together with fairly low chemical reactivity, make transition metals extremely useful. Examples include iron (and its alloy steel) for vehicle bodies and to reinforce concrete, copper for water pipes, and titanium for jet engine parts that must withstand high temperatures.

10.1.1 Electronic configurations in the d-block elements

The first d-series begins after calcium and contains the elements between scandium and zinc. Their highest energy electrons are in 3d orbitals. Figure 10.1 shows the energies of orbitals and their distances from the nucleus.

The 3d orbital is slightly higher in energy than the 4s orbital, and is lower in energy than the 4p orbital. But, it is closer to the nucleus than the 4s orbital. So after calcium, which has the electronic arrangement: $1s^2 2s^2 2p^6 3s^2 3p^6 4s^2$, the next ten electrons will go into the 3d orbitals, gradually filling this *inner* shell. With a few exceptions, the two outer electrons will be in 4s. This explains why the d-block elements are so similar.

Figure 10.2 shows the electron arrangements for the elements in the first d-series.

The arrangements of chromium, Cr, and copper, Cu, do not quite fit the pattern. 4s and 3d are very close in energy and electrons can easily move from one to another. Shells that are full ($3d^{10}$) or half full ($3d^5$) are particularly stable. This is rather like the extra stability of full electron shells in the noble gases.

10.1.2 Electronic configurations of the ions of d-block elements

To work out the configuration of the ion of an element, first write down the configuration of the element from its atomic number.

For example, to find the electron configuration of V^{2+}

Vanadium, V has an atomic number of 23. Its electron configuration is:

$1s^2 2s^2 2p^6 3s^2 3p^6 3d^3 4s^2$. The vanadium ion V^{2+} has lost the two $4s^2$ electrons and has the electron configuration: $1s^2 2s^2 2p^6 3s^2 3p^6 3d^3$.

Fig 10.1 *The energies of orbitals, showing their distances from the nucleus. The electron arrangement of calcium is shown*

Energy

Distance from nucleus

In fact with all **transition elements**, the 4s electrons are lost first because they are of the highest energy.

For example to find the electron configuration of the Cu^+ ion

The atomic number of copper is 29. The electron configuration is therefore $1s^2 2s^2 2p^6 3s^2 3p^6 3d^{10} 4s^1$. The Cu^+ ion has lost one electron so it has the electron configuration $1s^2 2s^2 2p^6 3s^2 3p^6 3d^{10}$.

10.1.3 The definition of a transition element

You may see the term transition elements used for elements in the d-block of the Periodic Table. But, the formal definition of a transition element is that it is one that forms at least one stable ion with a *part* full d-shell of electrons. Scandium only forms Sc^{3+}, which is $3d^0$, in all its compounds, and zinc only forms Zn^{2+} ($3d^{10}$) in all its compounds. They are therefore d-block elements but not transition elements.

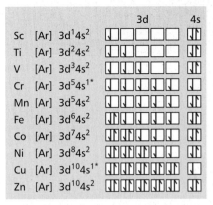

Fig 10.2 *Electronic arrangements of the elements in the first d-series. [Ar] represents the electron arrangement of argon – $1s^2 2s^2 2p^6 3s^2 3p^6$*

10.1.4 Chemical properties of transition elements

The chemistry of transition elements has four main features which are common to all the elements.

- **Variable oxidation states**. Transition metals have more than one oxidation state in their compounds, for example Cu(I) and Cu(II). They can therefore take part in many redox reactions.
- **Colour**. The majority of transition metal ions are coloured, for example Cu^{2+} is blue.
- **Catalysis**. Catalysts affect the rate of reaction without being chemically changed themselves. Many transition metals and their compounds show catalytic activity. For example, iron is the catalyst in the Haber process, vanadium(V) oxide in the Contact process and manganese(IV) oxide in the decomposition of hydrogen peroxide.
- **Complex formation**. d-block elements form **complex ions**. A complex ion is formed when a transition metal ion is surrounded by other molecules, which are bonded to it by dative bonds. For example, $[Cu(H_2O)_6]^{2+}$ is a complex ion that is formed when copper sulphate dissolves in water.

QUICK QUESTIONS

1 The electron arrangement of manganese is: $1s^2 2s^2 2p^6 3s^2 3p^6 3d^5 4s^2$
   a Write the electron arrangement of
      (i) a Mn^{2+} ion   (ii) a Mn^{3+} ion
   b Suggest why Mn^{2+} would be expected to be more stable.

2 The electron arrangement of iron can be written [Ar] $3d^6 4s^2$. What does the [Ar] represent?

3 a Write an equation for the formation of Sc^{3+} from Sc.
   b How many electrons are lost?
   c From which orbitals do these electrons come?

4 a Write an equation for the formation of Zn^{2+} from Zn.
   b How many electrons are lost?
   c From which orbitals do these electrons come?

Properties of transition elements

10.2.1 Variable oxidation states of transition elements

Group 1 metals lose their outer electron to form only 1+ ions and Group 2 lose their outer two electrons to form only 2+ ions in their compounds. A typical transition metal can use its 3d as well as its 4s electrons in bonding and this means that it can have a greater variety of oxidation states in different compounds. Table 10.1 shows this for the first transition series.

The most common oxidation states are shown in red, though they are not all stable.

Except for scandium and zinc (which are not transition metals) all the elements show the +I and +II oxidation states. These are formed by using only the 4s electrons for bonding.

For example, nickel has the electron configuration $1s^2 2s^2 2p^6 3s^2 3p^6 3d^8 4s^2$ and Ni^{2+} is $1s^2 2s^2 2p^6 3s^2 3p^6 3d^8$.

Iron has the electron configuration $1s^2 2s^2 2p^6 3s^2 3p^6 3d^6 4s^2$ and Fe^{2+} is $1s^2 2s^2 2p^6 3s^2 3p^6 3d^6$.

Note that only the lower oxidation states of transition metals actually exist as free ions, so that, for example, Mn^{2+} ions exist but Mn^{7+} ions do not. In all Mn(VII) compounds, as in MnO_4^-, manganese is covalently bonded, see Figure 10.3.

Fig 10.3 Bonding in the $[MnO_4]^-$ ion

Some examples of transition metal compounds – they are often coloured

Table 10.1 Oxidation numbers shown by the elements of the first d-series in their compounds

Sc	Ti	V	Cr	Mn	Fe	Co	Ni	Cu	Zn
	+I	+I	+I	+I	+I	+I	+I	+I	
	+II	+II	+II	+II	+II	+II	+II	+II	+II
+III	+III	+III	+III	+III	+III	+III	+III	+III	
	+IV	+IV	+IV	+IV	+IV	+IV	+IV		
		+V	+V	+V	+V	+V			
			+VI	+VI	+VI				
				+VII					

10.2.2 Colour of transition metal compounds

Many transition elements have coloured compounds, see Table 10.2. This is the result of spaces in the d-orbitals. These orbitals are not of exactly the same energy except in isolated gaseous atoms. So, electrons can move from one orbital to another of higher energy. In doing so they absorb electromagnetic energy of frequency in the visible region. For example, if a substance absorbs green light, it will let through red and blue and thus appear purple, see Topic 10.5.

10.2.3 Catalytic activity

Many transition elements and/or their compounds behave as catalysts.

Heterogeneous catalysts

Heterogeneous catalysts adsorb reactants on their surfaces by forming weak chemical bonds, see *Essential AS Chemistry for OCR*, Topic 14.3. This has two effects – weakening bonds within the reactant and holding the reactants close together on the metal surface in the correct orientation for reaction, see Figure 10.4. d-block metals have partly full d-orbitals which can be used to form bonds with adsorbed reactants. In this way, both transition metals and their compounds may act as heterogeneous catalysts.

Table 10.2 The colours of some transition metal ions

Ion	Colour	Ion	Colour
Ti^{3+}	purple	Fe^{3+}	brown
V^{2+}	purple	Co^{2+}	pink
Cr^{2+}	blue	Ni^{2+}	green
Mn^{2+}	pink	Cu^{2+}	blue

HINT

Adsorption is not the same as absorption. Adsorption involves the formation of weak bonds, such as van der Waals bonds, between the reactant and the surface.

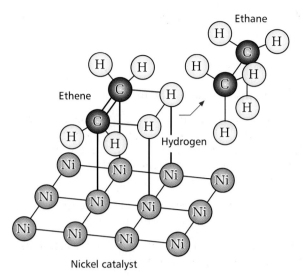

Fig 10.4 *Ethene and hydrogen adsorbed onto a nickel catalyst in the right orientation for new bonds (in red) to form*

Homogeneous catalysts

Transition metals compounds, with their variable oxidation states, can take part in redox (electron transfer) reactions. They do this by acting as a temporary 'store' for electrons. Suppose in a reaction between A and B, A transfers electrons to B. In the presence of a transition metal catalyst, A may pass electrons to the transition metal which in turn passes them on to reactant B. This indirect reaction may have a lower activation energy and therefore be faster than the direct reaction between A and B. Examples of d-block elements and compounds used as catalysts are shown in Table 10.3.

10.2.4 The formation of complex ions

Transition metals can form bonds by accepting electron pairs from other ions or molecules. The bonds that are formed are dative covalent bonds. An ion or molecule with a lone pair of electrons that forms a bond with a transition metal in this way is called a **ligand**. Some examples of ligands are: $H_2O:$, $:NH_3$, $:Cl^-$, $:CN^-$.

Often four or six ligands bond to a single transition metal ion and the resulting species is called a complex ion. The

number of ligands that surround the d-block metal ion is called the **coordination number**.

- Ions with coordination number six are usually octahedral, for example $[Co(NH_3)_6]^{3+}$.
- Ions with coordination number 4 are usually tetrahedral, for example, $[CoCl_4]^{2-}$.

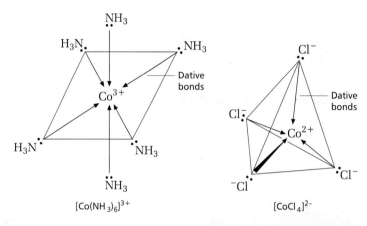

$[Co(NH_3)_6]^{3+}$ $[CoCl_4]^{2-}$

The $[Co(NH_3)_6]^{3+}$ ion, with six ligands, is an octahedron. Note that an octahedral shape has six points but *eight* faces. The metal ion, Co^{3+}, has a charge of +3 and as the ligands are all neutral, the complex ion has an overall charge of +3.

The $[CoCl_4]^{2-}$ ion, with four ligands, is a tetrahedron. The metal ion, Co^{2+} has a charge of +2 and each of the four ligands $:Cl^-$, has a charge of −1, so the complex ion has an overall charge of −2.

Complex ions may have a positive charge, negative charge or, less commonly, are neutral. Topics 10.3 and 10.4 give more detail of the properties of transition metals, using copper and iron as examples.

Table 10.3 *Some examples of catalysis involving d-block metals or their compounds*

Reaction	Catalyst
$N_2(g) + 3H_2(g) \rightarrow 2NH_3(g)$	Many metals, including Fe
$SO_2(g) + \frac{1}{2}O_2(g) \rightarrow SO_3(g)$	V_2O_5
$H_2O_2(aq) \rightarrow H_2O(l) + \frac{1}{2}O_2(g)$	MnO_2 + other metal oxides

QUICK QUESTIONS

1 a Explain the difference between a homogeneous and a heterogeneous catalyst.
  b Classify each of the examples below as homogeneous or heterogeneous.
    (i) A gauze of platinum and rhodium catalyses the oxidation of ammonia gas to nitrogen monoxide during the manufacture of nitric acid.
    (ii) Nickel catalyses the hydrogenation of vegetable oils.
    (iii) $Fe^{3+}(aq)$ ions catalyse the oxidation of iodide ions to iodine by persulphate ions in aqueous solution.

2 Which of the following could *not* be a ligand?
  H^+ $:NH_3$ O_2 $:Cl^-$

3 Explain what is meant by a dative covalent bond.

Some chemistry of copper and its compounds

Copper is one of the most used metals in the world – the annual word production is some 12 million tonnes. This is because it is relatively unreactive (so that it does not corrode) and it is a good conductor of both heat and electricity. Its uses include electrical wiring, plumbing pipes, coinage and heat exchangers.

Copper metal has the electron arrangement $1s^2\ 2s^2\ 2p^6\ 3s^2\ 3p^6\ 3d^{10}\ 4s^1$.

10.3.1 Cu$^+$ and Cu^{2+}

In most of its compounds, copper exists in oxidation state +II as the Cu^{2+} ion, which has the electron arrangement $1s^2\ 2s^2\ 2p^6\ 3s^2\ 3p^6\ 3d^9$. It has one space in the 3d orbital making it a transition metal. In some compounds copper exists in oxidation state +I, forming Cu^+, which is $3d^{10}$, so these compounds does not show properties, such as colour, which are typical of transition metal compounds.

10.3.2 The colour of copper compounds

The aqueous Cu^{2+} ion is blue. Most compounds containing Cu^+ are white, although the most familiar, copper(I) oxide, is brick red.

10.3.3 Precipitation reactions

The Cu^{2+} ion can be precipitated from solution by the addition of OH^- ions (for example as sodium hydroxide solution). The copper reacts to form a precipitate of pale blue, insoluble copper(II) hydroxide (see photo).

$$Cu^{2+}(aq) + 2OH^-(aq) \rightarrow Cu(OH)_2(s)$$

10.3.4 Ligand substitution reactions

The $Cu^{2+}(aq)$ ion exists in solution as the complex ion $[Cu(OH_2)_6]^{2+}$. The water molecules use their lone pairs to form dative covalent bonds with the copper ion. We say that the copper ion is complexed by six water molecules acting as ligands. This complex ion is responsible for the colour of copper sulphate solutions. It has an octahedral shape, see Figure 10.5. It has an overall 2+ charge (one Cu^{2+} and six neutral water molecules).

Water molecules (H_2O:) are not particularly good ligands and can be replaced by better ligands such as ammonia ($:NH_3$) or chloride ions ($:Cl^-$).

Displacement by chloride ions
Chloride is a better ligand than water because it is a negatively charged ion and is attracted electrostatically to the Cu^{2+}. So it can displace the water molecules. However, the chloride ion is bigger than a water molecule, and only four chloride ions can fit comfortably around a copper ion. The resulting complex ion has the formula $[CuCl_4]^{2-}$ and is yellowy-green.

The $[CuCl_4]^{2-}$ ion is tetrahedral, see Figure 10.6 and has an overall negative charge (one Cu^{2+} and four Cl^-).

Concentrated hydrochloric acid has a high concentration of chloride ions. If it is added a drop at a time to copper sulphate solution a gradual colour change from pale blue to yellowy-green takes place, as the chloride ligands replace water.

Displacement by ammonia
Ammonia is also a better ligand than water and will displace water from $[Cu(OH_2)_6]^{2+}$. This is because nitrogen, being less electronegative than oxygen,

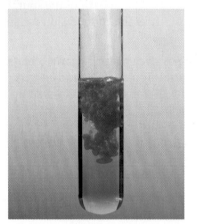

Uses of copper

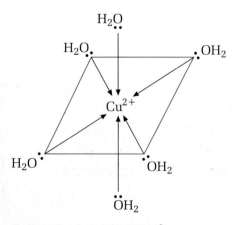

The pale blue precipitate of copper hydroxide

Fig 10.5 Octahedral $[Cu(OH_2)_6]^{2+}$

has a weaker hold on its lone pair of electrons than has oxygen and will more readily donate it to the copper ion.

In this case, ammonia displaces four of the water molecules resulting in the ion $[Cu(NH_3)_4(OH_2)_2]^{2+}$. This is octahedral, as it has six ligands surrounding the copper in the centre, see Figure 10.7. Notice that the four ammonia molecules are in the same plane with water molecules above and below. Ammonia and water are of comparable size and so the total number of ligands around the copper ions is not affected by the substitution of ammonia for water. The $[Cu(NH_3)_4(OH_2)_2]^{2+}$ ion is a striking deep blue colour.

Ammonia solution is alkaline and contains OH^- ions. When ammonia solution is added a drop at a time to a pale blue aqueous solution of copper sulphate the following changes take place.

At first the OH^- ions react with the copper ions to produce a precipitate of pale blue copper hydroxide – see above. When more ammonia is added, the copper hydroxide re-dissolves to form the deep blue $[Cu(NH_3)_4(OH_2)_2]^{2+}$ ion.

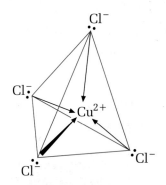

Fig 10.6 Tetrahedral $[CuCl_4]^{2-}$

Deep blue $[Cu(NH_3)_4(OH_2)_2]^{2+}$ ion

10.3.5 Catalytic activity

Like most transition metals, copper and its compounds can act as catalysts.

The reaction of zinc with acid
The reaction of zinc with hydrochloric acid or sulphuric acid is often used in the laboratory to produce hydrogen. Addition of a little copper sulphate solution speeds up the reaction. The copper sulphate first reacts with the zinc in a displacement reaction to form finely divided metallic copper, which is thought to be the catalyst rather than copper sulphate itself.

Oxidising propanone
If a heated strip of copper is placed in the vapour above a beaker of propanone, the copper begins to glow red-hot showing that an exothermic reaction is taking place on the surface. The copper is acting as a catalyst for the oxidation of propanone by oxygen in the air, to carbon dioxide and water.

This second example is a laboratory version of the use of catalysts, containing copper, to speed up the oxidation of organic pollutants (in exhaust systems, for example).

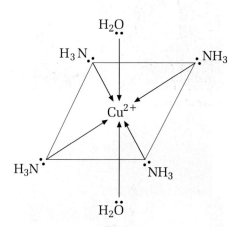

Fig 10.7 Octahedral $[Cu(NH_3)_4(OH_2)_2]^{2+}$

QUICK QUESTIONS

1 What is the H_2O—Cu—OH_2 bond angle in $[Cu(OH_2)_6]^{2+}$?

2 What is the Cl—Cu—Cl bond angle in $[CuCl_4]^{2-}$?

3 What is meant by the term catalyst? Give two points in your answer.

4 Explain why the chloride ion is significantly bigger than molecules of either water or ammonia.

5 a Write in the oxidation numbers of each element before and after the reaction:
$$Cu^{2+}(aq) + 2OH^-(aq) \rightarrow Cu(OH)_2(s)$$

b Is this a redox reaction? Explain your answer.

Iron is by far the most used metal in the world. Total production is around 560 million tonnes per year – most of this being converted into steel by alloying it with carbon and small amounts of other elements. The properties of the resulting steel can be tailored to a particular purpose because they depend on the type and amounts of the other elements added. Iron and steel are used for everything from food cans to car bodies to building bridges. Iron is not the ideal constructional material – it is quite dense and it rusts – but its abundance and cheapness more than make up for this.

Iron metal has the electron arrangement $1s^2\ 2s^2\ 2p^6\ 3s^2\ 3p^6\ 3d^6\ 4s^2$.

10.4.1 Redox reactions

Iron shows two common oxidation states in its compounds; $+II$ and $+III$. In the $+II$ state the two 4s electrons are lost when the Fe^{2+} ion is formed. In the $+III$ state both the 4s and one of the 3d electrons are lost to form the Fe^{3+} ion. The Fe^{3+} ion has the outer electron arrangement $3d^5$, which makes it especially stable – see Topic 10.1.1.

10.4.2 The colour of iron compounds

Compounds containing the Fe^{3+} ion are reddish brown in colour – like rust which is mostly iron(III) hydroxide. Compounds containing the Fe^{2+} ion are pale green in colour.

10.4.3 Precipitation reactions

Both the Fe^{2+} ion and the Fe^{3+} ion can be precipitated from solution by adding sodium hydroxide solution. The OH^- ions react with the iron ions to form insoluble hydroxides:

$$Fe^{2+}(aq) + 2OH^-(aq) \rightarrow Fe(OH)_2(s)$$
$$Fe^{3+}(aq) + 3OH^-(aq) \rightarrow Fe(OH)_3(s)$$

Iron(II) hydroxide is green and iron(III) hydroxide brown (see photo). Both colours are much darker than the colours of solutions of these ions and so are much easier to see. The colour of the precipitate formed on adding sodium hydroxide is therefore used as a test for the presence of these two ions.

Precipitate of iron(II) hydroxide (left) and iron(III) hydroxide (right)

Uses of iron

10.4.4 Ligand substitution reactions

In aqueous solution, both Fe^{3+} and Fe^{2+} exist as complex ions surrounded by six water molecules acting as ligands – $[Fe(H_2O)_6]^{3+}$ and $[Fe(H_2O)_6]^{2+}$, respectively. Both complexes are octahedral:

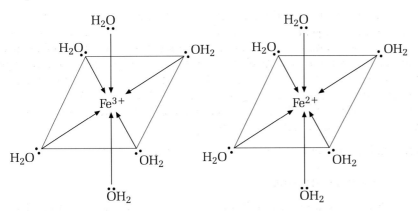

Good ligands can displace one or more of the water molecules in a ligand substitution reaction. One reaction of particular importance is that with the thiocyanate ion, SCN^-.

Displacement by thiocyanate

If potassium thiocyanate is added to a solution of $Fe^{3+}(aq)$, *one* of the water molecules is replaced by a thiocyanate ion:

$$[Fe(H_2O)_6]^{3+}(aq) + SCN^- \rightarrow [Fe(SCN)(H_2O)_5]^{2+}(aq) + H_2O(l)$$

The new complex formed has an intense blood-red colour that is easily recognisable and so this reaction can be used as a test for the $Fe^{3+}(aq)$ ion. It can also be used to measure its concentration. If excess potassium thiocyanate is used, all the $Fe^{3+}(aq)$ ions will be converted to the blood red complex and a colorimeter can be used to measure the intensity of the colour. The darker the colour, the greater the concentration of $Fe^{3+}(aq)$ ions.

10.4.5 Iron as a catalyst

One important use of iron as a catalyst is the Haber process for making ammonia from nitrogen and hydrogen (see *Essential AS Chemistry for OCR*, Topic 15.3). This is an example of heterogeneous catalyst – the iron is in pea-sized lumps, which gives it a larger surface area. The reactants are in the gas phase. It is a surface reaction in which nitrogen and hydrogen are adsorbed on the surface of the iron catalyst, where they react, and the ammonia produced is released from the surface.

QUICK QUESTIONS

1 a Write the electron configuration for (i) Fe^{2+}, (ii) Fe^{3+}.

  b Which is likely to be more stable? Explain your answer.

2 Give a reason why the ion SCN^- might be expected to be a better ligand than the water molecule.

3 In the reaction:

$$Fe^{3+}(aq) + 3OH^-(aq) \rightarrow Fe(OH)_3(s)$$

  a Put in the oxidation numbers of all the atoms before and after the reaction.

  b Is this a redox reaction? Explain your answer.

4 Explain what is meant by 'excess' potassium thiocyanate.

Most transition metal compounds are coloured. The colour is caused by the compounds absorbing energy that corresponds to light in the visible region of the spectrum. If a solution of a substance is purple it is because it absorbs green light from a beam of white light shone at it. The red and blue light pass through and the solution appears purple, see Figure 10.8.

Transition metal compounds are coloured because they have part-filled d-orbitals and it is possible for electrons to move from one d-orbital to another. In a separate atom, all the d-orbitals are of exactly the same energy, but in a compound, the presence of other atoms nearby makes the d-orbitals have different energies, so that electrons absorb energy in the visible region of the spectrum when they move from one to another.

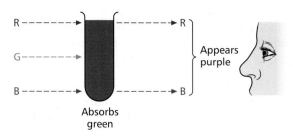

Fig 10.8 *Solutions look coloured because they absorb some colours and let others pass through*

10.5.1 The spectrophotometer

A spectrophotometer, Figure 10.9, is a special colorimeter that will produce an ultraviolet/visible spectrum. This is a graph of the amount of energy absorbed by a solution (called its 'absorbance') at different wavelengths (or frequencies).

In a spectrophotometer, a beam of white light is shone through the sample and the light that passes through hits a bank of detectors. The detectors measure how much light passes through the solution at each wavelength and the instrument plots this as a graph of absorbance (related to how much light is absorbed) against wavelength (or frequency). Most instruments look at the visible part of the spectrum and also into the ultraviolet.

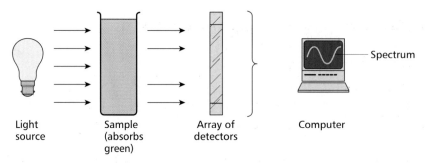

Fig 10.9 *A schematic diagram of a spectrophotometer*

The ultraviolet/visible spectrum

The spectrum of a typical transition metal ion, $[Ni(H_2O_6)]^{2+}(aq)$, is shown in Figure 10.10.

Where the graph line is high, the solution is absorbing light, where it is low, light is passing through. This solution absorbs blue and red light but lets through green and so looks green, see Figure 10.11.

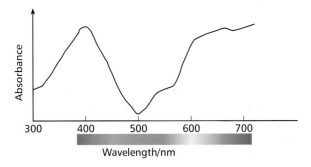

Fig 10.10 The ultraviolet/visible spectrum of a solution of [Ni(H₂O₆)]²⁺(aq)

Fig 10.11 A solution of Ni²⁺(aq)

10.5.2 Finding the formula of a transition metal complex using colorimetry

We can use a colorimeter to find the ratio of metal ions to ligands in a complex, which gives us the formula of the complex. We mix two solutions, one containing the metal ion and one the ligand, in different proportions. When they are mixed in the same ratio as they are in the complex, there is the maximum concentration of complex in the solution. So, the solution will absorb most light.

We will use the blood-red complex formed between Fe^{3+} ions and thiocyanate ions (SCN^-) as an example. We met with this reaction in Section 10.4.4. As the concentration of the red complex increases, less and less light will pass through the solution.

We usually make the experiment more sensitive by using a coloured filter in the colorimeter. We choose the filter by finding out the colour of light that the red solution absorbs most. Red absorbs light in the blue region of the visible spectrum, so we use a blue filter, Figure 10.12, so that only blue light passes into the sample tube.

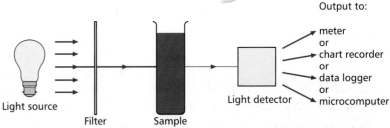

Fig 10.12 Using a colorimeter to find a formula

We start with two solutions of the same concentration, one containing Fe^{3+}(aq) ions, for example iron(III) sulphate, and one containing SCN^-(aq) ions, for example potassium thiocyanate. We mix them in the proportions shown in Table 10.4, adding water so that all the tubes have the same total volume of solution.

Table 10.4 The absorbance of different mixtures of Fe³⁺(aq) and SCN⁻(aq)

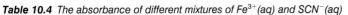

Tube	1	2	3	4	5	6	7	8
Vol. of Fe³⁺(aq) soln/cm³	10	10	10	10	10	10	10	10
Vol. of SCN⁻(aq) soln/cm³	2	4	6	8	10	12	14	16
Vol. of water/cm³	28	26	24	22	20	28	16	14
Absorbance	0.15	0.30	0.42	0.57	0.70	0.70	0.70	0.70

Each tube is put in the colorimeter and a reading of absorbance taken. A graph of absorbance is plotted against tube number, see Figure 10.13.

From the graph, the maximum absorbance occurs in tube 5; after this, adding more thiocyanate ions makes no difference. So this is the ratio of SCN^- ions to Fe^{3+} ions in the complex. From Table 10.4, tube 5 has equal amounts of SCN^- ions and Fe^{3+} ions so their ratio in the complex must be 1:1.

The formula is in fact $[FeSCN(H_2O)_5]^{2+}$, because the SCN^- has substituted for one of the water molecules in the complex ion $[Fe(H_2O)_6]^{3+}$, see Section 10.4.4.

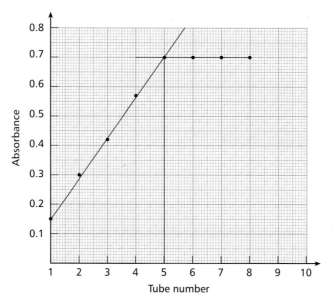

Fig 10.13 *A graph of absorbance against tube number*

QUICK QUESTIONS

1 Look at the ultraviolet/visible spectrum of a transition metal ion below.

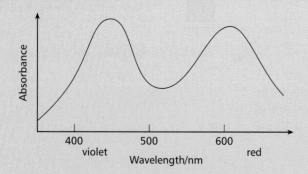

What colour will this ion appear to be? Explain your answer.

2 Scandium and zinc are found in the d-block of the Periodic Table. Write the electron arrangements of the ions Sc^{3+} and Zn^{2+}. Why are neither of these ions coloured?

3 A solution of the compound potassium manganate(VII) appears purple. What colours of light pass through this solution? What colours are absorbed by it? Make a rough sketch of its ultra violet/visible spectrum.

4 The graph below shows the absorbance of a series of mixtures containing different proportions of two solutions of the same concentration – one containing Ni^{2+} ions and the other containing a ligand called EDTA for short. The two solutions react together to form a coloured complex.

a Which mixture absorbs most light?

b Which mixture contains most complex?

c What is the simplest (empirical) formula of the complex?

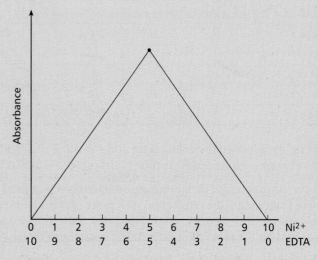

Redox reactions in transition metal chemistry

We saw in Section 10.2.1 that transition metals show more than one **oxidation state** in their compounds. Many of the reactions of transition metal compounds are redox reactions, in which the metals are either oxidised or reduced. Iron, for example, shows two stable oxidation states, Fe^{3+} and Fe^{2+}.

Fe^{2+} is the less stable state; it can be oxidised to Fe^{3+} by the oxygen in the air and also by chlorine. For example:

$$2Fe^{2+}(aq) + Cl_2(g) \rightarrow 2Fe^{3+}(aq) + 2Cl^-(aq)$$

In this reaction, chlorine is the **oxidising agent** – its oxidation number drops from 0 to –I (as it gains an electron) while that of the iron increases from +II to + III (as it loses an electron). Remember the phrase OIL RIG, oxidation is loss, reduction is gain (of electrons).

10.6.1 Half equations

We could build up the equation above by starting with two separate equations, one showing what is happening to the iron and one what is happening to the chlorine.

These are called **half equations**:

$$Fe^{2+}(aq) \rightarrow Fe^{3+}(aq) + e^- \quad Fe^{2+} \text{ is oxidised – it loses an electron}$$

and

$$Cl_2(aq) + 2e^- \rightarrow 2Cl^-(aq) \quad Cl_2 \text{ is reduced – it gains electrons}$$

Notice that electrons are on the opposite side of the arrow in the two half equations.

We now can add the two half equations to get the complete equation but only if there are the same number of electrons involved in each half equation. (We may need to multiply one or both of the half equations to make sure the number of electrons is the same in each case.)

So, we must multiply all the terms in the half equation for Fe^{2+} by 2 (shown in red) so that both half equations have the same numbers of electrons. Then we can add them together:

$$2Fe^{2+}(aq) \rightarrow 2Fe^{3+}(aq) + 2e^-$$
$$Cl_2(aq) + 2e^- \rightarrow 2Cl^-(aq)$$

$$\overline{2Fe^{2+}(aq) + Cl_2(aq) + \cancel{2e^-} \rightarrow 2Fe^{3+}(aq) + 2Cl^-(aq) + \cancel{2e^-}}$$

If we have multiplied correctly, there will be the same number of electrons on both sides of the arrow and they will cancel out leaving a balanced equation.

The technique of using half equations is useful for constructing balanced equations in more complex reactions. For example, potassium manganate(VII) can act as an oxidising agent in acidic solution (one containing $H^+(aq)$ ions). During the reaction the oxidation number of the manganese falls from +VII to +II. The half equation is:

$$MnO_4^-(aq) + 5e^- + 8H^+(aq) \rightarrow Mn^{2+}(aq) + 4H_2O(l)$$

Potassium manganate(VII) will also oxidise Fe^{2+} to Fe^{3+}. To construct a balanced symbol equation for the reaction of acidified potassium manganate (VII) with

$Fe^{2+}(aq)$ we must first multiply the Fe^{2+}/Fe^{3+} half reaction by 5 (so that the numbers of electrons in each half reaction are the same) and then add the two half equations:

$$5Fe^{2+}(aq) \rightarrow 5Fe^{3+}(aq) + 5e^-$$
$$MnO_4^-(aq) + 5e^- + 8H^+(aq) \rightarrow Mn^{2+}(aq) + 4H_2O(l)$$

$$5Fe^{2+}(aq) + MnO_4^-(aq) + \cancel{5e^-} + 8H^+(aq) \rightarrow 5Fe^{3+}(aq) + \cancel{5e^-} + Mn^{2+}(aq) + 4H_2O(l)$$

This technique can make balancing complex redox reactions much easier.

10.6.2 Redox titrations

We may wish to measure the concentration of an oxidising or a reducing agent. One way of doing this is to do a redox titration. This is similar in principle to an acid–base titration in which we find out how much acid is required to react with a certain volume of base (or *vice versa*).

One example is in the analysis of 'iron tablets' for quality control purposes. 'Iron tablets' contain iron(II) sulphate and may be taken by patients whose diet is short of iron for some reason.

As we have seen above, $Fe^{2+}(aq)$ reacts with manganate(VII) ions (in potassium manganate(VII)) in the ratio 5:1. The reaction does not need an indicator, because the colour of the mixture changes as the reaction proceeds, see Table 10.5.

Using a burette, we gradually add potassium manganate(VII) solution (the $MnO_4^-(aq)$ ions) to a solution containing acidified $Fe^{2+}(aq)$ ions. The purple colour disappears as the MnO_4^- ions are converted to pale pink Mn^{2+} ions to leave a virtually colourless solution. Once we have added just enough $MnO_4^-(aq)$ ions to react with all the $Fe^{2+}(aq)$ ions, one more drop of $MnO_4^-(aq)$ ions will turn the solution purple. This is the end point of the titration.

The apparatus used is shown in Figure 10.14.

Table 10.5 *The colours of the ions in the reaction between potassium manganate(VII) and iron(II) sulphate*

Ion	Colour
$Fe^{2+}(aq)$	Pale green
$MnO_4^-(aq)$	Intense purple
$Fe^{3+}(aq)$	Pale reddish brown
$Mn^{2+}(aq)$	Pale pink

NOTE

The body needs iron compounds to make haemoglobin, the compound that carries oxygen in the blood.

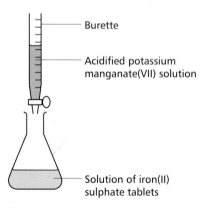

Burette

Acidified potassium manganate(VII) solution

Solution of iron(II) sulphate tablets

Fig 10.14 *Apparatus for a titration*

Example
A brand of iron tablets stated on the pack 'Each tablet contains 0.200 g of iron(II) sulphate'. The following experiment was done to check this.

One tablet was dissolved in excess sulphuric acid and made up to 250 cm³ in a volumetric flask. 25.00 cm³ of this solution was pipetted into a flask and titrated with 0.005 mol dm⁻³ potassium manganate(VII) solution until the solution just became purple. Taking an average of several titrations, 5.26 cm³ of potassium manganate(VII) solution was needed.

$$\text{Number of moles potassium manganate(VII) solution} = \frac{M \times V}{1000}$$

where M is the concentration of the solution in mol dm⁻³ and V is the volume of solution used in cm³

Number of moles potassium manganate(VII) solution $= 0.005 \times \dfrac{5.26}{1000}$

$$= 2.63 \times 10^{-5} \text{ mol}$$

$$5Fe^{2+}(aq) + MnO_4^-(aq) + 8H^+(aq) \rightarrow 5Fe^{3+}(aq) + Mn^{2+}(aq) + 4H_2O(l)$$

From the equation, 5 mol of Fe^{2+} react with 1 mol of MnO_4^-.

Number of moles of $Fe^{2+} = 5 \times 2.63 \times 10^{-5}$ mol $= 1.315 \times 10^{-4}$ mol

25.00 cm^3 of solution contained $\frac{1}{10}$ tablet

So 1 tablet contains $1.315 \times 10^{-4} \times 10 = 1.315 \times 10^{-3}$ mol Fe^{2+}

Since 1 mol iron(II) sulphate contains 1 mol Fe^{2+}, each tablet contains 1.315×10^{-3} mol $FeSO_4$

The relative formula mass of $FeSO_4$ is 151.9

So, each tablet contains $1.315 \times 10^{-3} \times 151.9 = 0.200$ g of iron(II) sulphate as stated on the bottle

QUICK QUESTIONS

1 In the blast furnace, iron is produced by the following reaction:

$$Fe_2O_3 + 3C \rightarrow 2Fe + 3CO$$

   a Give the oxidation numbers of iron, carbon and oxygen (i) before, and (ii) after the reaction.

   b What is the reducing agent?

   c What is the oxidising agent?

2 Zinc will reduce V^{3+} ions to V^{2+} ions. The two half equations are:

   $Zn(s) \rightarrow Zn^{2+}(aq) + 2e^-$

   $V^{3+}(aq) + e^- \rightarrow V^{2+}(aq)$

   Write the balanced equation for this.

3 A titration to determine the amount of iron(II)sulphate in an iron tablet was carried out. The tablet was dissolved in excess sulphuric acid and made up to 250 cm^3 in a volumetric flask. 25.00 cm^3 of this solution was pipetted into a flask and titrated with 0.005 mol dm^{-3} potassium manganate(VII) solution until the solution just became purple. Taking an average of several titrations, 5.00 cm^3 of potassium manganate(VII) solution was needed. How many grams of iron are in this tablet? A_r Fe = 55.8.

1 The table below relates to oxides of Period 3 in the Periodic Table.

Oxide	Na_2O	MgO	Al_2O_3	SiO_2	P_4O_{10}	SO_3
Melting point/°C	1275	2827	2017	1607	580	33
Bonding						
Structure						

a Complete the table using the following guidelines.
   (i) Complete the 'bonding' row using **only** the words: *ionic* or *covalent*.
   (ii) Complete the 'structure' row using **only** the words: *simple molecular* or *giant*.
   (iii) Explain, in terms of forces, the difference between the melting points of MgO and SO_3. [5]

b The oxides Na_2O and SO_3 were each added separately to water. For each oxide, construct a balanced equation for its reaction with water.
   (i) SO_3 reaction with water
   (ii) Na_2O reaction with water   [2]

[Total: 7]

OCR, Specimen 2000

2 The lattice enthalpy of rubidium chloride, RbCl, can be determined indirectly using a Born–Haber cycle.

a Use the data in the table below to construct the cycle to determine a value for the lattice enthalpy of rubidium chloride.

Enthalpy change	Energy/kJ mol^{-1}
Formation of rubidium chloride	−435
Atomisation of rubidium	+81
Atomisation of chlorine	+122
1st ionisation energy of rubidium	+403
1st electron affinity of chlorine	−349 [6]

b Explain why the lattice enthalpy of lithium chloride, LiCl, is more exothermic than that of rubidium chloride.   [2]

[Total: 8]

OCR, Specimen 2000

3 Aqueous copper(II) sulphate reacts with an excess of aqueous ammonia to give a dark blue solution. The solution contains the octahedral complex ion, $[Cu(NH_3)_x(H_2O)_y]^{2+}$.

The formula of this complex ion can be determined using colorimetry.

• A student makes up six different mixtures of $1.00\ mol\ dm^{-3}$ aqueous ammonia and $0.500\ mol\ dm^{-3}$ aqueous copper(II) sulphate and water.

• She filters the mixtures to remove any precipitate that forms.
• The filtrate is a dark blue solution that contains the complex ion, $[Cu(NH_3)_x(H_2O)_y]^{2+}$.
• The student places the blue solution into a container and measures the absorbance of the solution.

The table below shows the relative absorbance of each mixture.

Mixture	one	two	three	four	five	six
Volume of $0.500\ mol\ dm^{-3}$ $CuSO_4(aq)/cm^3$	5.00	5.00	5.00	5.00	5.00	5.00
Volume of $1.00\ mol\ dm^{-3}$ $NH_3(aq)/cm^3$	3.00	6.00	9.00	11.00	15.00	18.00
Volume of $H_2O(l)/cm^3$	17.00	14.00	11.00	9.00	5.00	2.00
Relative absorbance	0.29	0.57	0.86	0.95	0.94	0.95

a Copper is a transition element. One typical property of a transition element is that it forms coloured complex ions.

State **two** other typical properties of a transition element.   [2]

b The precipitate formed when the student makes some of the mixtures is copper(II) hydroxide.
   (i) Write an ionic equation to show the formation of copper(II) hydroxide from its ions.   [1]
   (ii) If this precipitate is **not** removed, an inaccurate absorbance reading is obtained. Suggest why.   [1]

c Draw a graph of the relative absorbance against the volume of aqueous ammonia using the grid below.

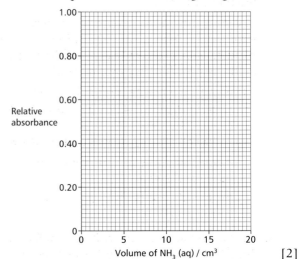

[2]

d (i) How many moles of copper(II) sulphate are there in $5.00\ cm^3$ of a $0.500\ mol\ dm^{-3}$ solution?   [1]

(ii) Use the graph to estimate the **smallest** volume of 1.00 mol dm^{-3} aqueous ammonia that gives the maximum relative absorbance. [1]

(iii) How many moles of ammonia are there in the volume in (ii)? [1]

(iv) Deduce the values of x and y in the formula of the octahedral complex ion, $[Cu(NH_3)_x(H_2O)_y]^{2+}$. [1]

e In the octahedral complex, $[Cu(NH_3)_x(H_2O)_y]^{2+}$ ammonia is a ligand.

(i) Explain why ammonia can behave as a ligand. [1]

(ii) The bond angle around the nitrogen atom in an ammonia molecule is $107°$ but it is $109.5°$ in the octahedral complex. Explain why the bond angles differ. [2]

f Aqueous copper(II) ions react with concentrated hydrochloric acid to give a yellow solution of $[CuCl_4]^{2-}(aq)$. This reaction is an example of ligand substitution.

(i) Write an equation to show the formation of $[CuCl_4]^{2-}(aq)$. [1]

(ii) Draw the shape of the $[CuCl_4]^{2-}$ ion. [1]

[Total: 15]

OCR, June 2003

4 Iron is a typical transition element.

- Iron shows more than one oxidation state in its compounds.
- Iron and its compounds are used as catalysts.

a Complete the electronic configuration for an **iron(III) ion, Fe^{3+}**, and use it to explain why iron is a transition element.
Fe^{3+}: $1s^2 2s^2 2p^6$ [2]

b State **one** use of iron or one of its compounds as a catalyst. State the name of the catalyst and the reaction catalysed. [1]

c Under certain conditions iron can be oxidised to form sodium ferrate, Na_2FeO_4. This is a red-purple coloured substance that has properties very similar to that of potassium manganate(VII).

(i) Analysis of a sample of sodium ferrate showed that it contains the following percentage composition by mass,

Na, 27.74%, Fe, 33.66% and O, 38.60%.

Show that these data are consistent with the formula Na_2FeO_4. [2]

(ii) Deduce the oxidation state of iron in sodium ferrate, Na_2FeO_4. [1]

d The half-equation for the reduction of ferrate ions, FeO_4^{2-}, in acidic conditions is shown below.

$$FeO_4^{2-} + 8H^+ + 4e^- \rightarrow Fe^{2+} + 4H_2O$$

Acidified $FeO_4^{2-}(aq)$ ions oxidise aqueous iodide ions, I^-, to form aqueous iodine, I_2.

(i) Construct the half-equation for the oxidation of iodide ions to form iodine. [1]

(ii) Construct the ionic equation for the redox reaction that occurs between aqueous FeO_4^{2-} and aqueous I^- in the presence of H^+. [2]

(iii) Predict the colour change you would see when aqueous FeO_4^{2-} is added to an excess of aqueous I^- in the presence of H^+. [1]

[Total: 10]

OCR, Jan 2003

5 a (i) Explain what is meant by the term *transition element*. [1]

(ii) Complete the electronic configuration of the vanadium atom.
$1s^2 2s^2 2p^6$ [1]

b Aqueous transition metal ions can react with aqueous hydroxide ions.

(i) Complete the table below.

Metal ion	**Formula** and **state symbol** of the product of the reaction with $OH^-(aq)$	Colour of product
$Fe^{2+}(aq)$		
$Fe^{3+}(aq)$		

[5]

(ii) Aqueous ammonia reacts with water in the following way.

$$NH_3(aq) + H_2O(l) \rightleftharpoons NH_4^+(aq) + OH^-(aq)$$

When aqueous ammonia is added dropwise to aqueous copper(II) ions, a very pale blue precipitate is observed which disappears in excess ammonia to give a deep blue solution.

Write equations to show the formation from aqueous copper(II) ions of the pale blue precipitate and the deep blue solution. [4]

[Total: 11]

OCR, June 2002

6 The manganate(VII) ion, MnO_4^-, is a strong oxidising agent frequently used in laboratory analysis. It reacts with the ethanedioate ion, $C_2O_4^{2-}$, in hot acidic solution to form CO_2 and M^{2+} ions.

$$MnO_4^- + 8H^+ + 5e^- \rightarrow Mn^{2+} + 4H_2O$$
$$C_2O_4^{2-} \rightarrow 2CO_2 + 2e^-$$

a Construct the full ionic equation for this reaction. [2]

b Calculate the volume of $0.0200 \text{ mol dm}^{-3}$ potassium manganate(VII) required to react with 25.0 cm^3 of $0.0400 \text{ mol dm}^{-3}$ sodium ethanedioate. [3]

[Total: 5]

OCR, June 2002

11

Unifying concepts

How fast?

11.1

The rate of chemical reactions

The main factors that affect the rate of chemical reactions are: temperature, concentration, pressure, surface area and catalysts, see *Essential AS Chemistry for OCR*, Chapter 14. In this topic we will look at the measurement of reaction rates.

11.1.1 What is a reaction rate?

As the reaction: $A + 2B \rightarrow C$ takes place, the concentrations of the reactants A and B decrease with time and that of the product C increases with time. We could follow the concentration of A, B or C with time and plot the results as shown in Figure 11.1.

The rate of the reaction is defined as the change in concentration (of any of the reactants or products) with time. But notice how different the graphs are for A, B and C. For this reason it is important to state whether we are following A, B or C. We usually assume that a rate is measured by following the concentration of a product C, because the concentration of the product increases with time.

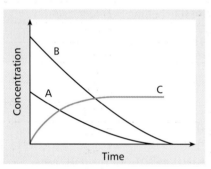

Fig 11.1 *Changes of concentration with time for A, B and C*

11.1.2 The average rate of a reaction

We can find an average rate from a graph of concentration against time, see Figure 11.2, for example. The average rate of the reaction with respect to C, during a period of time Δt is the change in concentration of C, $\Delta[C]$ divided by Δt.

The average rate of reaction between t_1 and $t_2 = \dfrac{\Delta[C]}{\Delta t}$.

If in 10 seconds $[C]$ changed from 1.0 mol dm^{-3} to 1.1 dm^{-3} then:

$$\text{Average rate of reaction in this period} = \frac{1.1 - 1.0 \, \text{mol dm}^{-3}}{10 \, \text{s}}$$

$$= \frac{0.1}{10} = 1 \times 10^{-2} \, \text{mol dm}^{-3} \, \text{s}^{-1}$$

> **HINT**
>
> Square brackets around a chemical symbol, [ ], are used to indicate its concentration in mol dm^{-3}.

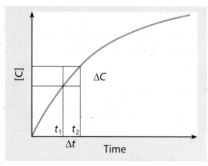

Figure 11.2 *The average rate of reaction between time t_1 and $t_2 = \dfrac{\Delta[C]}{\Delta t}$*

> **HINT**
>
> Remember the use of the symbol Δ (delta) to mean a change in a quantity.

Notice that the units of this rate of reaction are mol dm^{-3} s^{-1}

11.1.3 The rate of reaction at any instant

We are often interested in the rate *at a particular instant* in time rather than the *average* rate of change over a period of time. To find the rate of change of $[C]$ at a particular instant we draw a tangent to the curve at that time and then find its gradient (slope), Figure 11.3.

11.1.4 Measuring a reaction rate

To measure a reaction rate, we need a method of measuring the concentration of one of the reactants or products over a period of time (keeping the temperature constant, because rate varies with temperature). The method we choose will depend on the substance whose concentration is being measured and also the speed of the reaction. For example, in the reaction between bromine and methanoic acid the solution starts off being brown (from the presence of bromine) and ends up being colourless.

$$Br_2(aq) + HCO_2H(aq) \rightarrow 2Br^-(aq) + 2H^+(aq) + CO_2(g)$$

So, we can use a colorimeter to measure the decreasing concentration of Br_2. Table 11.1 shows some typical results.

In order to find the reaction rate at different times, we plot the results on a graph and then measure the gradients of the tangents at the times required, Figure 11.4. For example when $t = 0$, 5 and 10 minutes.

At $t = 0$, rate of reaction = 0.010/4.0 = 0.0025 mol dm^{-3} min^{-1}

At $t = 5$, rate of reaction = 0.0076/9.0 = 0.00073 mol dm^{-3} min^{-1}

At $t = 10$, rate of reaction = 0.0046/14 = 0.00033 mol dm^{-3} min^{-1}

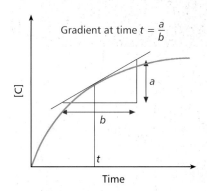

Fig 11.3 The rate of change of [C] at time t, is the gradient of the concentration–time graph at t

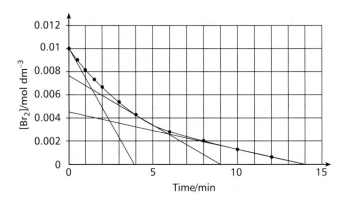

Fig 11.4 Finding the rate of reaction at t = 0, t = 5 and t = 10 for bromine reacting with methanoic acid

Table 11.1 [Br$_2$] measured over time

Time/min	[Br$_2$]/mol dm^{-3}
0.0	0.0100
0.5	0.0090
1.0	0.0081
1.5	0.0073
2.0	0.0066
3.0	0.0053
4.0	0.0042
6.0	0.0027
8.0	0.0020
10.0	0.0013
12.0	0.0007

QUICK QUESTIONS

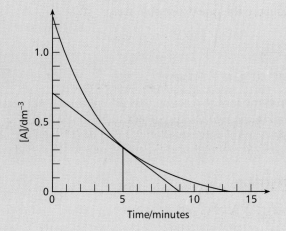

1 Is the concentration being plotted that of a reactant or a product? Explain your answer.

2 The tangent to the curve at time = 5 minutes is drawn on the graph. Find the gradient of the tangent – remember to include units.

3 What does the gradient represent?

4 Without drawing tangents, what can be said about the gradients of the tangents drawn at time = 0 and time = 10 minutes.

5 Explain your answer to question 4.

Answer the following questions about the reaction rate graph above.

NOTE

Species is a general term that includes molecules, ions and atoms that might be involved in a chemical reaction.

The rate of a chemical reaction depends on the concentrations of some or all of the species in the reaction vessel – reactants, products and catalysts. But these do not necessarily all make the same contribution to how fast the reaction goes. The **rate expression** tells us about the contribution of the species that do affect the reaction rate.

For example, in the reaction $X + Y \rightarrow Z$, the concentration of X, [X], may have more effect than the concentration of Y, [Y]. Or, it may be that [X] has no effect on the rate and only [Y] matters, or even that [Z] has an effect on the rate. The detail of how each species contributes to the rate of the reaction can only be found out by experiment.

11.2.1 The rate expression

The rate expression is the result of experimental investigation. It is an equation that describes how the rate of the reaction depends on the concentration of species involved in the reaction. It is quite possible that one (or more) of the species that appear in the chemical equation will not appear in the rate expression. This means that they do not affect the rate. For example, the reaction:

$$X + Y \rightarrow Z$$

might have the rate expression:

Rate $\propto$ [X][Y], where the symbol $\propto$ means 'proportional to'.

NOTE

If $A \propto B$ then if A is doubled, then B also doubles.

This would mean that both [X] and [Y] have an equal effect on the rate. Doubling either [X] or [Y] would double the rate of the reaction. Doubling the concentration of both would quadruple the rate.

But it might be that the rate expression for the reaction is:

$$\text{Rate} \propto [X][Y]^2$$

This would mean that doubling [X] would double the rate of the reaction, but doubling [Y] would quadruple the rate.

11.2.2 The rate constant *k*

We can get rid of the proportionality sign if we introduce a constant to this expression. For example, suppose the rate expression were:

$$\text{Rate} \propto [X][Y]^2$$

This can be written:

$$\text{Rate} = k[X][Y]^2$$

and *k* is called the **rate constant** for the reaction. *k* is different for every reaction and varies with temperature, so the temperature needs to be stated. If the concentration of all the species in the rate equation is 1 mol dm^{-3}, then the rate of reaction is *k*.

NOTE

Remember that the rate expression is entirely derived from experimental evidence and that it *cannot* be predicted from the chemical equation for the reaction. It is therefore quite unlike the equilibrium law expression that you will meet in Chapter 12, (although it looks similar to it at first sight).

11.2.3 The order of a reaction

Suppose the rate equation for a reaction is:

$$\text{Rate} = k[X][Y]^2$$

This means that [Y], which is raised to the power of 2, has double the effect on the rate than that of [X]. The **order of reaction**, with respect to one of the species, is the power to which the concentration of that species is raised in the rate equation. It tells us how the rate depends on the concentration of that species.

So, for Rate = $k[X][Y]^2$ the order with respect to X is one, ($[X]$ and $[X]^1$ are the same thing) and the order with respect to Y is two.

The overall order of the reaction is the sum of the orders of all the species, which appear in the rate expression. In this case the overall order is three. So this reaction is said to be first order with respect to X, second order with respect to Y and third order overall.

11.2.4 The chemical equation and the rate equation

The rate equation tells us about the species that affect the rate. Species that appear in the chemical equation do not necessarily appear in the rate equation. Also, the coefficient of a species in the chemical equation – the number in front of it – has no relevance to the rate expression. But catalysts, which do not appear in the chemical equation, *may* appear in the rate equation.

For example, in the reaction:

$$CH_3CO.CH_3(aq) + I_2(aq) \xrightarrow{\text{H}^+ \text{ catalyst}} CH_2ICOCH_3(aq) + HI(aq)$$
propanone   iodine   iodopropanone   hydrogen iodide

The rate equation has been found by experiment to be:

$$\text{Rate} = k[CH_3COCH_3(aq)][H^+(aq)]$$

So the reaction is first order with respect to propanone, first order with respect to H^+ ions and second order overall. The rate does not depend on $[I_2 (aq)]$, so we can say the reaction is zero order with respect to iodine, in other words the rate depends on $[I_2]^0$ The H^+ ions act as a catalyst in this reaction.

11.2.5 Units of the rate constant, *k*

The units of rate constant vary depending on the overall order of the reaction.

For a first order reaction where:

$$\text{Rate} = k[A]$$

the units of rate are mol dm^{-3} s^{-1} and the units of $[A]$ are mol dm^{-3} so the units of k are s^{-1} obtained by cancelling:

$$\cancel{\text{mol dm}^{-3}} \text{ s}^{-1} = k \cancel{\text{mol dm}^{-3}}$$

the unit of *k* for a first order reaction is s^{-1}.

For a second order reaction where:

$$\text{Rate} = k[B][C]$$

the units of rate are mol dm^{-3} s^{-1} = and the units of both $[A]$ and $[B]$ are mol dm^{-3} so by cancelling:

$$\cancel{\text{mol dm}^{-3}} \text{ s}^{-1} = k \cancel{\text{mol dm}^{-3}} \times \text{mol dm}^{-3}$$

the units of *k* for a second order reaction are dm^3 mol^{-1} s^{-1}.

QUICK QUESTIONS

1 Write down the rate expression for a reaction that is first order with respect to [A], first order with respect to [B] and second order with respect to [C].

2 For the reaction:

$$BrO_3^-(aq) + 5Br^-(aq) + 6H^+(aq)$$
bromate ions   bromide ions   hydrogen ions

$$\longrightarrow 3Br_2(aq) + 3H_2O(l)$$
bromine   water

the rate expression is Rate = $k[BrO_3^-(aq)][Br^-(aq)][H^+(aq)]^2$

a What is the order with respect to (i) $BrO_3^-(aq)$, (ii) $Br^-(aq)$, (iii) $H^+(aq)$?

b What would happen to the rate if we doubled the concentration of (i) $BrO_3^-(aq)$, (ii) $Br^-(aq)$, (iii) $H^+(aq)$?

c What are the coefficients of (i) $BrO_3^-(aq)$, (ii) $Br^-(aq)$, (iii) $H^+(aq)$, (iv) $Br_2(aq)$ and (v) $H_2O(l)$ in the chemical equation above?

d Work out the units for the rate constant.

3 In the reaction L + M → N the rate expression is found to be:

$$\text{Rate} = k[L]^2[H^+]$$

a What is k?

b What is the order of the reaction with respect to (i) L, (ii) M, (iii) N, (iv) H^+?

c What is the overall order of the reaction?

d The rate is measured in mol dm^{-3}. What are the units of k?

e Suggest the function of H^+ in the reaction.

4 In the reaction G + 2H → I + J, which is the correct rate expression?

a Rate = $k[G][H]^2$

b Rate = $k[G][H]/[I][J]$

c Rate = $k[G][H]$

d It is impossible to tell without experimental data.

HINT

It is best to work out the units rather than try to remember them.

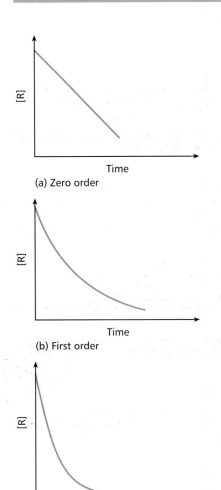

Fig 11.5 *Concentration–time graphs for zero-, first-, and second-order reactions*

The rate expression tells us how the rate of a reaction depends on the concentration of the species involved. It only includes the species that affect the rate of the reaction, see Topic 11.2.

- If the rate is not affected by the concentration of a species, the reaction is zero order with respect to that species. We do not include this species in the rate expression.
- If the rate is directly proportional to the concentration of the species, the reaction is first order with respect to that species.
- If the rate is proportional to square of the concentration of the species, the reaction is second order with respect to that species, and so on.

11.3.1 Finding the order of a reaction from a graph

The usual outcome from a reaction rate experiment will be a series of readings of concentration, at different times, of one of the species involved in the reaction. Usually, the first step in finding the order with respect to the species is to plot these readings on a concentration–time graph.

To help interpretation of the experiment, we make sure that the concentrations of all the reactants, other than the one being measured, stay the same. We do this by making their concentration so large compared with the reactant being investigated, that any change is negligible. For example:

$$R \quad + \quad B \quad \rightarrow \quad products$$
$$\text{reactant R} \quad \text{reactant B}$$

At the start of the reaction, we make [B] very large compared with [R]. For example:

$[R] = 0.01 \text{ mol dm}^{-3}$

$[B] = 1.00 \text{ mol dm}^{-3}$

At the end of the reaction:

$[R] = 0$ (all used up)

$[B] = 1.00 - 0.01 = 0.99 \text{ mol dm}^{-3}$

So during the whole reaction [B] is practically constant.

The graph of [R] plotted against time, might look like one of those in Figure 11.5.

Notice that in every case the concentration of R decreases with time. This is because R is a reactant and is therefore being used up.

It is often possible to tell the order with respect to the species measured simply by looking at the graph. If the graph is a straight line, as in Figure 11.5a, the gradient of the graph, which represents the rate, is the same whatever the concentration of R. The order with respect to R is therefore zero.

First and second order reactions both give curves but the second order curve is deeper than the first order one, see Figure 11.5 (b) and (c).

11.3.2 Finding the half-life of a first order reaction

The half-life of a reaction, $t_{1/2}$, is the time taken for [R] to fall from any chosen value to half that value. The first half-life is often taken from the start of the reaction. The second half-life is from the end of the first, and so on. Figure 11.6 is a concentration–time graph for a first order reaction and shows some successive half-lives.

HINT

If we had plotted the concentration of a *product* against time, its concentration would increase.

HINT

You will probably have met half-life in the context of radioactive decay.

In a first-order reaction all the half-lives are the same (within experimental error). The half-life is therefore independent of the concentration for a first order reaction. For a second (or greater) order reaction, successive half-lives increase.

11.3.3 Finding the order of a reaction by using rate–concentration graphs

A second method of determining the order of a reaction with respect to a particular species is by plotting a graph of rate against concentration (rather than concentration against time).

We start with the original graph of [R] against time, and draw tangents at different values of [R]. The gradients of these tangents are the reaction rates (the changes in concentration over time) at different concentrations, see Figure 11.7. The values for these rates can then be used to construct a second graph of rate against concentration, see Figure 11.8.

If the graph is a horizontal straight line (Figure 11.8a), this means that the rate is unaffected by [R] so the order is zero.

If this graph is a sloping straight line through the origin (Figure 11.8b) then rate $\propto$ [R]^1 so the order is 1.

If this graph is not a straight line (Figure 11.8c) we cannot find the order directly – it is probably 2. Try plotting rate against [R]^2. If this is a straight line then the order is 2.

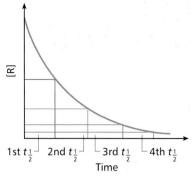

Fig 11.6 *Successive half-lives for a first-order reaction are all the same*

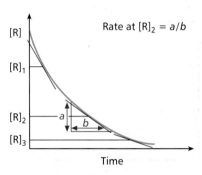

Fig 11.7 *Finding the rate of reaction at different values of [R]*

QUICK QUESTIONS

1 A particular reaction has the concentration–time graph shown:

   a Is F a reactant or a product? Explain your answer.

   b The order of the reaction with respect to F is either 0 or 1 or 2. Which of these can you definitely eliminate just by looking at the graph? Explain your answer.

   c If you suspect that the order of the reaction with respect to F is 1, how can you test the shape of the graph to confirm this? Say what you would do and the result you would expect if the order is 1.

2 A different reaction has the concentration–time graph shown.

   a The order of the reaction with respect to G is either 0 or 1 or 2. Which of these can you eliminate just by looking at the shape of the graph?

   b What is the expected pattern of half-lives for:
   (i) a first order reaction,
   (ii) a second order reaction?

   c By making a rough estimate of half-lives from the graph, what is the order of reaction with respect to G? Explain your answer.

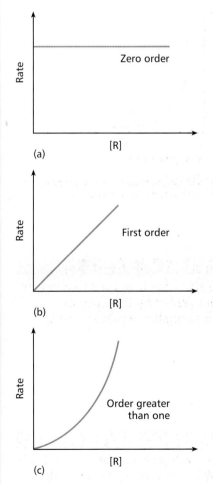

Fig 11.8 *Graphs of rate against concentration*

The initial rate method

The methods in Topic 11.3 only allow you to find the order of a reaction with respect to the *one* reactant whose concentration has been measured. The initial rate method allows you to find the order with respect to any species in the reaction mixture.

11.4.1 The initial rate experiments

With the initial rate method, a series of experiments is carried out. Each experiment starts with a different combination of initial concentrations of reactants, catalysts, etc. The experiments are planned so that, between any pair of experiments, there is only one concentration that varies – the rest stay the same. Then, for each experiment, the concentration of one reactant is followed and a concentration–time graph plotted, Figure 11.9. The tangent to the graph at time = 0 is drawn. The gradient of this tangent is the initial rate. The beauty of measuring the *initial* rate is that the concentrations of all substances in the reaction mixture are known *exactly* at this time.

Comparing the initial concentration and the initial rates for pairs of experiments allows the order with respect to each reactant to be found. For example, for the reaction:

$$2NO(g) \quad + \quad O_2(g) \quad \rightarrow \quad 2NO_2(g)$$

nitrogen monoxide oxygen nitrogen dioxide

The initial rates are shown in Table 11.2.

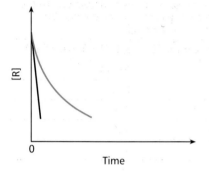

Fig 11.9 *Finding the initial rate of a reaction. The initial rate is the gradient at time = 0*

Table 11.2 *Results obtained for the reaction $2NO(g) + O_2(g) \rightarrow 2NO_2(g)$*

Experiment number	Initial [NO]/10^{-3} mol dm^{-3}	Initial [O_2]/10^{-3} mol dm^{-3}	Initial rate/10^{-4} mol dm^{-3} s^{-1}
1	1	1	7
2	2	1	28
3	3	1	63
4	2	2	56
5	3	3	189

Comparing experiment 1 with experiment 2, [NO] is doubled while [O_2] stays the same. The rate quadruples (from 7 to 28) which suggests rate $\propto$ [NO]^2. This is confirmed by comparing experiments 1 and 3 where [NO] is trebled in experiment 3 while [O_2] stays the same. Here, in experiment 3, the rate is increased ninefold as would be expected if rate $\propto$ [NO]^2. So the order with respect to NO is two.

Now compare experiment 2 with 4. Here [NO] is constant but [O_2] doubles. The rate doubles (from 28 to 56) so it looks as if rate $\propto$ [O_2]. This is confirmed by considering experiments 3 and 5. Again [NO] is constant, but [O_2] triples. The rate triples too, confirming that the order with respect to O_2 is 1.

So   rate $\propto$ [NO]^2

      rate $\propto$ [O_2]^1

i.e.   rate $\propto$ [NO]^2[O_2]^1.

Provided that no other species affect the reaction rate, the overall order is 3 and the rate expression is:

      rate $= k$ [NO]^2[O_2]^1

11.4.2 Finding the rate constant *k*

To find *k* in the equation above, we simply substitute any set of values of rate, [NO] and [O$_2$] in the equation.

Taking the values for experiment 2:

28×10^{-4} mol dm^{-3} s^{-1} = k $(2 \times 10^{-3})^2$ (mol dm^{-3}) (mol dm^{-3}) $\times 1 \times 10^{-3}$ mol dm^{-3}

Cancelling the units where we can:

28×10^{-4} ~~mol dm^{-3}~~ s^{-1} = k (4×10^{-6}) ~~(mol dm^{-3})~~ (mol dm^{-3}) $\times 1 \times 10^{-3}$ mol dm^{-3}

$$28 \times 10^{-4} = k \times 4 \times 10^{-9} \text{ mol}^2 \text{ dm}^{-6} \text{ s}^{-1}$$

$$k = \frac{28}{4} \times 10^5 \text{ dm}^6 \text{ mol}^{-2} \text{ s}^{-1}$$

$$k = 7 \times 10^5 \text{ dm}^6 \text{ mol}^{-2} \text{ s}^{-1}$$

Since the units of *k* vary for reactions of different orders, it is important to put the units for rate and the concentrations in and then cancel them to make sure you have the correct units for *k*.

11.4.3 The effect of temperature on *k*

Small changes in temperature produce large changes in reaction rates. A rule of thumb is that for every 10 K rise in temperature, the rate of a reaction doubles. Suppose the rate expression for a reaction is rate = *k* [A][B]. We know that [A] and [B] do not change with temperature, so the rate constant, *k*, must increase with temperature.

In fact, the rate constant, *k*, is a measure of the speed of a reaction. The larger the value of *k*, the faster the reaction. Look at Table 11.3. You can see that *k* goes up with temperature. This is true of all reactions.

Why the rate constant depends on temperature

We saw in *Essential AS Chemistry for OCR*, Topic 14.2, that temperature is a measure of the average speed of molecules. Particles will only react together if their collisions have enough energy to start bond breaking. This energy is called the **activation energy, E_A**. Figure 11.10 shows how the energies of the particles in a gas (or in a solution) are distributed at three different temperatures. Only molecules with energy greater than E_A can react.

The shape of the graph changes with temperature. As the temperature increases, a greater proportion of molecules has enough energy to react. This is the main reason for the increase in reaction rate with temperature.

Table 11.3 *The values of the rate constant, k, at different temperatures for the reaction* $2HI(g) \rightarrow I_2(g) + H_2(g)$

Temperature /K	$k/10^{-3}$ dm^{-3} mol^{-1} s^{-1}
633	0.0178
666	0.107
697	0.501
715	1.05
781	15.1

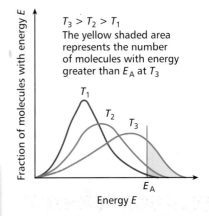

$T_3 > T_2 > T_1$
The yellow shaded area represents the number of molecules with energy greater than E_A at T_3

Fig 11.10 *The distribution of molecular energies at three temperatures*

QUICK QUESTIONS

1 For the reaction A + B → C, the following data were obtained:

Initial [A]	Initial [B]	Initial rate
1	1	3
1	2	12
2	2	24

a What is the order of reaction with respect to: (i) A, (ii) B?

b What is the overall order?

c What would be the initial rate if the initial [A] were 1 and [B] were 3?

d What do these results suggest is the rate expression for this reaction?

e Can we be certain that this is the full rate expression? Explain your answer.

The rate-determining step

Most reactions take place in more than one step. The separate steps that lead from reactants to products are together called the **reaction mechanism**. For example, the reaction below involves 12 ions:

$$BrO_3^-(aq) + 6H^+(aq) + 5Br^-(aq) \rightarrow 3Br_2(aq) + 3H_2O(l)$$

This reaction *must* take place in several steps – it is very unlikely indeed that the 12 ions of the reactants will all collide at the same time. The steps in between will involve very short-lived intermediates. These intermediate species, which would tell us about the mechanism of the reaction, are usually impossible to isolate and therefore identify. So, we must use other ways of working out the mechanism of the reaction.

11.5.1 The rate-determining step

In a multi-step reaction, the steps nearly always follow after each other, so that the product(s) of one step is/are the starting material(s) for the next. Therefore the rate of the slowest step governs the rate of the whole process. The slowest step forms a 'bottleneck', called the **rate-determining step**. Suppose you had everything you needed to make a cup of coffee, starting with cold water. The rate of getting your drink will be governed by the slowest step – waiting for the kettle to boil, no matter how quickly you get the cup out of the cupboard and the coffee out of the jar.

In a chemical reaction, any step that occurs *after* the rate-determining step will not affect the rate. So species that are involved in the mechanism after the rate-determining step will not appear in the rate expression. For example, the reaction:

$$A + B + C \rightarrow Y + Z$$

might occur in the following steps:

1. A + B $\xrightarrow{\text{fast}}$ D (1st intermediate)

2. D $\xrightarrow{\text{slow}}$ E (2nd intermediate)

3. E + C $\xrightarrow{\text{fast}}$ Y + Z

Step 2 is the slowest step and so determines the rate. Then, as soon as some E is produced, it rapidly reacts with C to produce Y and Z.

But the rate of step 1 might affect the overall rate – the concentration of D depends on this. So, any species involved in or *before* the rate-determining step could affect the rate and therefore appear in the rate expression.

The reaction between iodine and propanone shows this. (You will not be expected to recall details for an examination).

The overall reaction is:

$$CH_3-\overset{O}{\overset{||}{C}}-CH_3(aq) + I_2(aq) \xrightarrow[\text{catalyst}]{H^+} CH_2ICCH_3(aq) + HI(aq)$$

propanone   iodine

and the rate expression is found to be rate = k [CH$_3$OCCH$_3$][H$^+$]

The mechanism is:

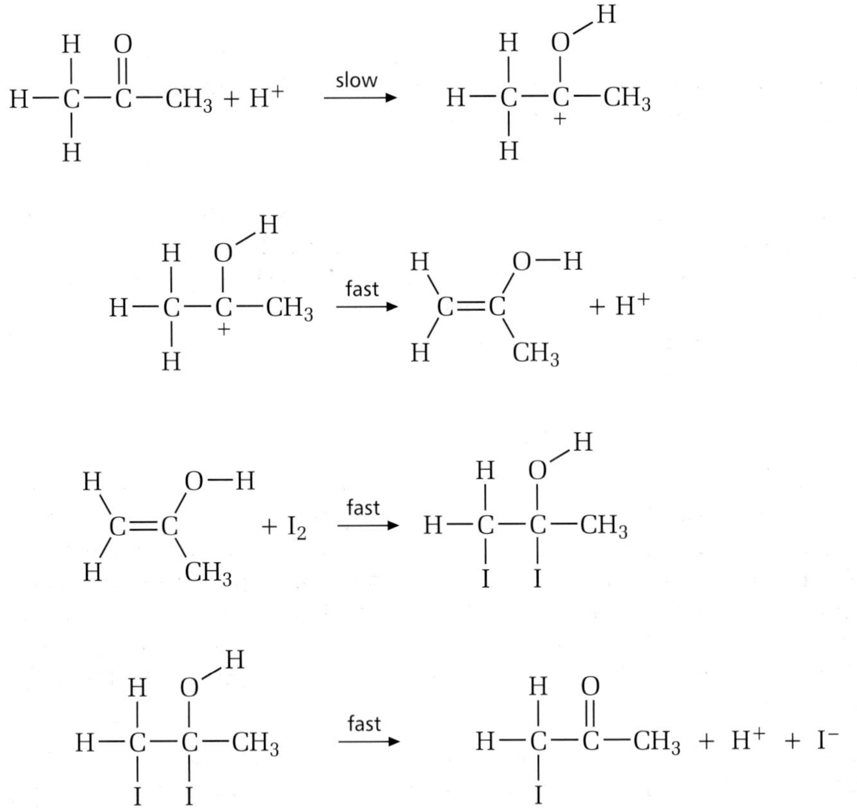

The rate-determining step is the first one, which explains why I$_2$ does *not* appear in the rate expression.

11.5.2 Using the rate-determining step to find a reaction mechanism

Sometimes working out the rate-determining step also helps to work out a reaction mechanism. Here is a simple example. The three structural isomers with formula C$_4$H$_9$Br, all react with alkali. The overall reaction is:

$$C_4H_9Br + OH^- \rightarrow C_4H_9OH + Br^-$$

Two mechanisms are possible.

(a) A two-step mechanism:

step 1
$$\overset{\text{slow}}{C_4H_9Br \rightarrow C_4H_9^+ + Br^-}$$

followed by, step 2 $$\overset{\text{fast}}{C_4H_9^+ + OH^- \rightarrow C_4H_9OH}$$

The slow step involves breaking the C—Br bond while the second (fast) step is a reaction between oppositely charged ions.

(b) A one-step mechanism:
$$C_4H_9Br + OH^- \rightarrow C_4H_9OH + Br^-$$

The C—Br bond breaks at the same time as the C—OH bond is forming.

The three isomers of formula C_4H_9Br are:

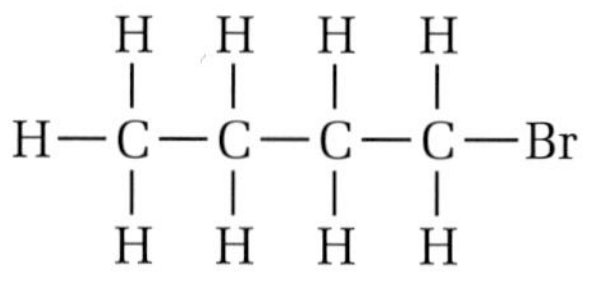

1-bromobutane

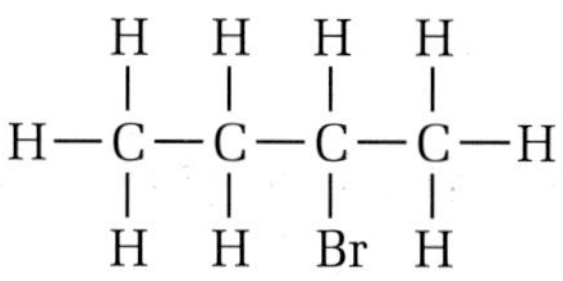

2-bromobutane

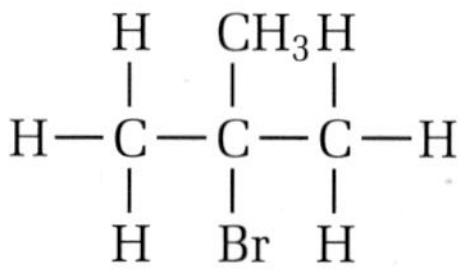

2-bromo-2-methylpropane

Experiments show that 1-bromobutane reacts by a second order mechanism: rate = k [C_4H_9Br] [OH^-]. The rate depends on the concentration of *both* the bromobutane *and* the OH^- ions, suggesting mechanism (b).

Experiments show that 2-bromo-2-methylpropane reacts by a first order mechanism: rate = k [C_4H_9Br]. This suggests mechanism (a).

2-bromobutane reacts by a mixture of both mechanisms and has a more complex rate expression.

QUICK QUESTIONS

1 Baking a cake involves the following steps
   a weighing the ingredients,
   b mixing the ingredients,
   c baking in the oven.
   Which is the rate-determining step?

2 The following reaction schemes show possible mechanisms for the overall reaction:

catalyst
A + E → G

Scheme 1	Scheme 2	Scheme 3
(i) slow A + B → C	(i) fast A + B → C	(i) fast A + B → C
(ii) fast C → D + B	(ii) fast C → D + B	(ii) slow C → D + B
(iii) fast D + E → F	(iii) slow D + E → F	(iii) fast D + E → F
(iv) fast F → G	(iv) slow F → G	(iv) fast F → G

   a In scheme 2, which species is the catalyst?
   b Which species *cannot* appear in the rate expression for scheme 1?
   c Which is the rate determining step in scheme 3?

How far?

Chemical equilibrium

We saw in *Essential AS Chemistry for OCR*, Chapter 15, that many reactions are reversible and do not go to completion, but instead end up as an equilibrium mixture of reactant and products. A reversible reaction that can reach equilibrium is denoted by the symbol $\rightleftharpoons$. In this topic we see how we can tackle equilibrium reactions mathematically.

12.1.1 Equilibrium reactions

Many organic reactions are reversible and will reach equilibrium in time. The reaction between ethanol, C_2H_5OH, and ethanoic acid, CH_3CO_2H, to produce the ester ethyl ethanoate, $CH_3CO_2C_2H_5$, and water is typical.

If ethanol and ethanoic acid are mixed in a flask (stoppered to prevent evaporation) and left, a mixture is eventually obtained in which *all four* substances are present. (A strong acid catalyst is required if this is to occur within a reasonable length of time). The system has reached equilibrium and we can write:

$$C_2H_5OH\ (aq) + CH_3CO_2H\ (aq) \rightleftharpoons CH_3CO_2C_2H_5(aq) + H_2O(l)$$

ethanol   ethanoic acid   ethyl ethanoate   water

Titrating the ethanoic acid to investigate the equilibrium position

The mixture may be analysed by titrating the ethanoic acid with standard alkali (allowing for the catalyst). This gives the number of moles of ethanoic acid. From this we can work the number of moles of the other components, (and from this their concentration, if the total volume of the mixture is known).

HINT

The concentration of a solution is the number of moles of solute dissolved in $1\ dm^{-3}$ of solution. A square bracket around a formula is shorthand for 'concentration of that substance in $mol\ dm^{-3}$'.

If several experiments are done with different quantities of starting materials, it is always found that the ratio:

$$\frac{[CH_3CO_2C_2H_5(aq)]_{eqm}[H_2O(aq)]_{eqm}}{[CH_3CO_2H(aq)]_{eqm}[C_2H_5OH(aq)]_{eqm}}$$

has a constant value, provided the experiments are done at the same temperature. This constant is called the equilibrium constant and has the symbol K_c. The subscript 'eqm' means that the concentrations have been measured when equilibrium has been reached. The subscript 'c' stands for concentration – the equilibrium constant is a ratio of concentrations.

12.1.2 The equilibrium law and the equilibrium constant, K_c

The expression above is an example of a general law – the equilibrium law. This is expressed as follows. For a reaction:

$$a\text{A} + b\text{B} + c\text{C} \rightleftharpoons x\text{X} + y\text{Y} + z\text{Z}$$

the expression $\dfrac{[X]_{eqm}{}^{X}[Y]_{eqm}{}^{Y}[Z]_{eqm}{}^{Z}}{[A]_{eqm}{}^{A}[B]_{eqm}{}^{B}[C]_{eqm}{}^{C}}$

has a constant value, K_c, provided the temperature is constant. K_c is called the equilibrium constant and is different for different reactions. It changes with temperature. The units of K_c vary, and you must work them out for each reaction by cancelling out the units of each term, for example:

$$A + B \rightleftharpoons C \qquad K_c = \dfrac{[C]^1}{[A]^1[B]^1}$$

$$\text{Units are } \dfrac{\cancel{\text{mol dm}^{-3}}}{\cancel{\text{mol dm}^{-3}} - x \text{ mol dm}^{-3}} = \dfrac{1}{\text{mol dm}^{-3}}$$

$$= \text{dm}^3 \text{ mol}^{-1}$$

K_c is found by experiment for any particular reaction at a given temperature.

To find K_c for the reaction between ethanol and ethanoic acid:

0.10 mol of ethanol is mixed with 0.10 mol of a solution of ethanoic acid and allowed to reach equilibrium. The total volume of the system is 20.0 cm^3 (0.020 dm^3). We find by titration that 0.033 mol ethanoic acid is present once equilibrium is reached.

From this we can work out the number of moles of the other components present at equilibrium:

At start

$$\begin{array}{cccc} C_2H_5OH\,(l) + CH_3CO_2H\,(l) &\rightleftharpoons& CH_3CO_2C_2H_5(l) + H_2O(l) \\ \text{0.10 mol} \quad\quad \text{0.10 mol} && \text{0 mol} \quad\quad\quad \text{0 mol} \end{array}$$

We know that there are 0.033 mol of CH_3CO_2H at equilibrium. This means:

- that there must also be 0.033 mol of C_2H_5OH at equilibrium. (The equation tells us that they react 1:1 and we know we started with the same number of moles of each.)
- that $(0.10 - 0.033) = 0.067$ moles of CH_3CO_2H have been used up. The equation tells us that when 1 mole of CH_3CO_2H is used up, 1 mole each of $CH_3CO_2C_2H_5$ and H_2O are produced. So, there must be 0.067 moles of each of these.

At equilibrium

$$\begin{array}{cccc} C_2H_5OH\,(l) + CH_3CO_2H\,(l) &\rightleftharpoons& CH_3CO_2C_2H_5(l) + H_2O(l) \\ \text{0.033 mol} \quad\quad \text{0.033 mol} && \text{0.067 mol} \quad\quad \text{0.067 mol} \end{array}$$

We need the *concentrations* of the components at equilibrium. As the volume of the system is 0.20 dm^3 these are:

$$\begin{array}{cccc} C_2H_5OH\,(l) &+& CH_3CO_2H\,(l) &\rightleftharpoons& CH_3CO_2C_2H_5(l) &+& H_2O(l) \\ \text{0.033/0.020 mol dm}^{-3} && \text{0.033/0.020 mol dm}^{-3} && \text{0.067/0.020 mol dm}^{-3} && \text{0.067/0.020 mol dm}^{-3} \end{array}$$

We enter the concentrations into the equilibrium equation:

$$K_c = \dfrac{[CH_3CO_2C_2H_5(aq)_{eqm}][H_2O(aq)]_{eqm}}{[CH_3CO_2H(aq)]_{eqm}[C_2H_5OH(aq)]_{eqm}}$$

$$K_c = \dfrac{[0.067/0.020 \text{ mol dm}^{-3}]_{eqm}[0.067/0.020 \text{ ml dm}^{-3}]_{eqm}}{[0.033/0.020 \text{ mol dm}^{-3}]_{eqm}[0.033/0.020 \text{ mol dm}^{-3}]_{eqm}} = 4.1$$

The units all cancel out, and the volumes (0.020 dm^3) cancel out, so in this case we didn't need to know the volume of the system, so $K_c = 4.1$. It has no unit.

12.1.3 K_c and the position of equilibrium

The size of the equilibrium constant, K_c, can tell us about the composition of the equilibrium mixture. The equilibrium law expression is always of the general form $\dfrac{[\text{products}]}{[\text{reactants}]}$. So:

- If K_c is much greater than 1, products predominate over reactants. We usually say that the equilibrium is over to the right.
- If K_c is much less than 1, reactants predominate, and the equilibrium position is over to the left.

Reactions where the equilibrium constant is greater than 10^{10} are usually regarded as going to completion while those with an equilibrium constant of less than 10^{-10} are regarded as not taking place at all.

QUICK QUESTIONS

1 Write down the equilibrium law expressions for the following:

  a $A + B \rightleftharpoons C$

  b $2A + B \rightleftharpoons C$

  c $2A + 2B \rightleftharpoons 2C$

2 Work out the units for K_c for question 1 **a** to **c**.

3 For the reaction above between ethanol and ethanoic acid, at a different temperature, the equilibrium mixture was found to contain 0.117 mol of ethanoic acid, 0.017 mol of ethanol, 0.083 mol ethyl ethanoate and 0.083 mol of water.

  a Calculate K_c.

  b Why do you not need to know the volume of the system to calculate K_c?

  c Is the equilibrium further to the right or to the left compared with the one above?

12.2 Calculations using the equilibrium law expressions

12.2.1 Calculating the composition of a reaction mixture

We can use the equilibrium law to calculate the composition of a reaction mixture that has reached equilibrium.

Example 1

The reaction of ethanol and ethanoic acid is:

$$C_2H_5OH\ (l) + CH_3CO_2H\ (l) \rightleftharpoons CH_3CO_2C_2H_5(l) + H_2O(l)$$

ethanol ethanoic acid ethyl ethanoate water

We know that:

$$K_c = \frac{[CH_3CO_2C_2H_5(aq)]_{eqm}[H_2O(aq)]_{eqm}}{[CH_3CO_2H(aq)]_{eqm}[C_2H_5OH(aq)]_{eqm}}$$

Suppose that $K_c = 4$ at the temperature of our experiment and we want to know how much ethyl ethanoate we could produce by mixing 1 mol of ethanol and 1 mol of ethanoic acid. Set out the information as shown below:

Equation: $C_2H_5OH\ (l) + CH_3CO_2H\ (l) \rightleftharpoons CH_3CO_2C_2H_5(l) + H_2O(l)$

	ethanol	ethanoic acid	ethyl ethanoate	water
At start:	1 mol	1 mol	0 mol	0 mol
At equilibrium:	(1 − x) mol	(1 − x) mol	x mol	x mol

We do not know how many moles of ethyl ethanoate will be produced, so we call this x. The equation tells us that x mol of water will also be produced and in doing so x mol of both ethanol and ethanoic acid will be used up. So the amount of each of these remaining at equilibrium is $(1 - x)$ mol.

These figures are in moles, but we need concentrations in mol dm^{-3} to substitute in the equilibrium law expression. Suppose the volume of the system at equilibrium was V dm^{-3}. Then:

$$[C_2H_5OH(aq)]_{eqm} = \frac{(1-x)}{V}\ \text{mol dm}^{-3}$$

$$[CH_3CO_2H(aq)]_{eqm} = \frac{(1-x)}{V}\ \text{mol dm}^{-3}$$

$$[CH_3CO_2C_2H_5(aq)]_{eqm} = \frac{x}{V}\ \text{mol dm}^{-3}$$

$$[H_2O(aq)]_{eqm} = \frac{x}{V}\ \text{mol dm}^{-3}$$

These figures may now be put into the expression for K_c:

$$K_c = \frac{x/\cancel{V} \times x/\cancel{V}}{(1-x)/\cancel{V} \times (1-x)/\cancel{V}}$$

The Vs cancel, so *in this case* we do not need to know the actual volume of the system.

$$4 = \frac{x \times x}{(1-x) \times (1-x)}$$

$$4 = \frac{x^2}{(1-x)^2}$$

Taking the square root of both sides, we get:

$$2 = \frac{x}{(1-x)}$$
$$2\ (1-x) = x$$

HINT

The volume of the reaction mixture will cancel for all systems with equal numbers of moles of products and reactants, so V is sometimes omitted. But, it is always better to include V and cancel it out later, so you will not forget it for systems where the Vs do not cancel out.

112

$$2 - 2x = x$$
$$2 = 3x$$
$$x = \tfrac{2}{3}$$

So $\tfrac{2}{3}$ mol of ethyl ethanoate and $\tfrac{2}{3}$ mol of water are produced if the reaction reaches equilibrium, and the composition of the equilibrium mixture would be: ethanol $\tfrac{1}{3}$ mol, ethanoic acid $\tfrac{1}{3}$ mol, ethyl ethanoate $\tfrac{2}{3}$ mol, water $\tfrac{2}{3}$ mol.

12.2.2 Calculating the amount of a reactant needed

We can also use K_c to find the amount of a reactant needed to give a required amount of product.

For the following reaction in ethanol solution, $K_c = 30.0$ dm^3 mol^{-1}:

$$CH_3COCH_3 \quad + \quad HCN \quad \rightleftharpoons \quad CH_3C(CN)(OH)CH_3$$

propanone   hydrogen cyanide   2-hydroxy-2-methylpropanenitrile

$$K_c = \frac{[CH_3C(CN)(OH)CH_3]_{eqm}}{[CH_3COCH_3]_{eqm}[HCN]_{eqm}} = 30.0 \text{ dm}^3 \text{ mol}^{-1}$$

Suppose we are carrying out this reaction in 2.00 dm^3 of ethanol. How much hydrogen cyanide is required to produce 1.00 mol of product if we start with 4.00 mol of propanone? Set out as before with the quantities at the start and at equilibrium.

At equilibrium, we want 1 mol of product. Let x be the number of moles of HCN required.

Equation:	CH_3COCH_3	+	HCN	$\rightleftharpoons$	$CH_3C(CN)(OH)CH_3$
At start:	4.00 mol		x mol		0 mol
At equilibrium:	(4.00 − 1.00) mol		(x − 1.00) mol		1.00 mol
	3.00 mol		(x − 1.00) mol		1.00 mol

These are the numbers of moles but we need the *concentrations* to put in the equilibrium law expression. The volume of the solution is 2.00 dm^3 and the units for concentration are mol dm^{-3} so we next divide each quantity by 2.00 dm^3.

So at equilibrium $[CH_3COCH_3]_{eqm} = 3.00/2.00$ mol dm^{-3}

$$[HCN]_{eqm} = (x - 1.00)/2.00 \text{ mol dm}^{-3}$$

$$[CH_3C(CN)(OH)CH_3]_{eqm} = 1.00/2.00 \text{ mol dm}^{-3}$$

Putting the figures into the equilibrium law expression:

$$30.0 \text{ dm}^3 \text{ mol}^{-1} = \frac{1.00/2.00 \text{ mol cm}^{-3}}{3.00/2.00 \text{ mol dm}^{-3} \times (x - 1.00)/2.00 \text{ mol cm}^{-3}}$$

Cancelling through and rearranging we have:

$$30(\tfrac{3}{2}(x-1)/2) = \tfrac{1}{2}$$
$$45(x - 1) = 1$$
$$45x = 46$$
$$x = \tfrac{46}{45} = 1.02$$

So, to obtain 1 mol of product we must start with 1.02 mol hydrogen cyanide, if the volume of the system is 2.00 dm^3.

In this example the volume of the system *does* make a difference, because this reaction does not have the same number of moles of products and reactants.

QUICK QUESTIONS

1 Try reworking the problem in Section 12.2.2 above with the same conditions but:

a with a volume of 1.00 dm^3 of ethanol.

b with a starting amount of 2 mol of propanone.

c to produce 2 mol of product.

Gaseous equilibria

Gases may react together to reach equilibrium and they obey the equilibrium law (see Topic 12.1), but we express the concentration of a gas in a different way, using partial pressures (rather than concentration in $mol\,dm^{-3}$).

12.3.1 Partial pressure

In a mixture of gases, each gas contributes its pressure to the total pressure of the gas mixture. The contribution of a gas is called the **partial pressure**.

- The partial pressure of any one gas is the pressure it would produce if it occupied the container on its own.
- The sum of all the partial pressures of the gases in the mixture adds up to the total pressure.
- The symbol p is used for partial pressure.

For example, air is a mixture of one-fifth oxygen and four-fifths nitrogen. If the total pressure is atmospheric, $100\,kPa$:

the partial pressure of oxygen, $pO_2 = \frac{1}{5} \times 100\,kPa = 20\,kPa$

the partial pressure of nitrogen, $pN_2 = \frac{4}{5} \times 100\,kPa = 80\,kPa$

which, when added together give the total pressure $100\,kPa$.

Pressures are usually measured in kilopascals, kPa.

12.3.2 Mole fractions

More precisely, the partial pressure of any gas in a mixture is given by its mole fraction multiplied by the total pressure.

> **The mole fraction of a gas A =** $\dfrac{\textbf{number of moles of gas A}}{\textbf{total number of moles of gases in the mixture}}$

The mole fraction of oxygen in air is $\frac{1}{5}$ and that of nitrogen $\frac{4}{5}$. (Remember that a mole of any gas has the same volume, so if we know the ratio of the volumes of gases, we know the ratio of moles.)

12.3.3 Applying the equilibrium law to gaseous equilibria

Equilibrium constants for gaseous reactions are given the symbol K_p. When dealing with gases we usually work in partial pressures.

For the reaction $\quad aA(g) + bB(g) + cC(g) \rightleftharpoons xX(g) + yY(g) + zZ(g)$

$$K_p = \frac{pX(g)_{eqm}{}^x\, pY(g)_{eqm}{}^y\, pZ(g)_{eqm}{}^z}{pA(g)_{eqm}{}^a\, p\,B(g)_{eqm}{}^b\, C(g)_{eqm}{}^c}$$

Notice the similarity with the equilibrium law expression for K_c given in Topic 12.1.

12.3.4 The units of K_p

As with K_c, K_p may or may not have units depending on the expression from which it derives. You will need to work out the units for each one.

Example 1
For the equilibrium:

$$H_2(g) + I_2(g) \rightleftharpoons 2HI(g)$$

$$K_p = \frac{pHI(g)_{eqm}^2}{pH_2(g)_{eqm}\,pI_2(g)_{eqm}} \quad \text{and the units are} \quad \frac{\cancel{kPa}^2}{\cancel{kPa} \times \cancel{kPa}}$$

This particular K_p has no units as they cancel.

Notice that the equation could equally well have been written:

$$2HI(g) \rightleftharpoons H_2(g) + I_2(g)$$

In which case:

$$K_p = \frac{pH_2(g)_{eqm}\,pI_2(g)_{eqm}}{pHI(g)_{eqm}^2}$$

and though the units still cancel, K_p will have a different value, so it is *vital* that you write the equation when you are working out equilibrium constants.

Example 2
This is the key step in the Haber process for the manufacture of ammonia.

$$3H_2(g) + N_2(g) \rightleftharpoons 2NH_3(g)$$

$$K_p = \frac{pH_2(g)_{eqm}^3\,pN_2(g)_{eqm}}{pNH_3(g)_{eqm}^2}$$

The units for this equilibrium constant are kPa^{-2}

An industrial plant that manufactures ammonia using the equilibrium reaction in Example 2

12.3.5 Using K_p to find partial pressures

K_p is 0.020 at 700 K for the reaction:

$$2HI(g) \rightleftharpoons H_2(g) + I_2(g)$$

If the reaction started with pure HI and the initial pressure of HI was 100.0 kPa, what will be the partial pressure of the hydrogen when equilibrium is reached?

Let the partial pressure of hydrogen at equilibrium $(pH_2(g)_{eqm})$ be x kPa.

The chemical equation tells us:

- that there will be the same number of moles of hydrogen and iodine at equilibrium, therefore if $pH_2(g)_{eqm} = x$, then $pI_2(g)_{eqm} = x$ as well.
- that for each mole of hydrogen or (iodine) produced, *two* moles of hydrogen iodide are *used up*, therefore if $pH_2(g)_{eqm}$ is x, $pHI(g)_{eqm}$ must be $(100 - 2x)$.

Set out the calculation in the same way as when using K_c, see Section 12.2.

	2HI(g)	$\rightleftharpoons$	H₂(g)	+	I₂(g)
At start:	100.0 kPa		0 kPa		0 kPa
At equilibrium:	(100.0 − 2x) kPa		x kPa		x kPa

$$K_p = \frac{pH_2(g)_{eqm}\,pI_2(g)_{eqm}}{pHI(g)^2}$$

Putting in the figures gives:

$$0.020 = \frac{x \times x}{(100.0 - 2x)^2} = \frac{x^2}{(100.0 - 2x)^2}$$

Taking the square root of each side gives:

$$0.141 = \frac{x}{100.0 - 2x}$$

$$0.141\,(100.0 - 2x) = x$$

$$14.1 - 0.282x = x$$

$$14.1 = 1.282x$$

$$x = \frac{14.1}{1.282}$$

$$x = 10.99$$

$$pH_2(g)_{eqm} = 10.99\,kPa = 11\,kPa \text{ (to 2 significant figures)}$$

QUICK QUESTIONS

1 In the example opposite, what will be the partial pressures of iodine and hydrogen iodide at equilibrium?

2 Write the expression for K_p for:

   a $2SO_2 + O_2 \rightleftharpoons 2SO_3$

   b $N_2O_4(g) \rightleftharpoons 2NO_2(g)$

   c $H_2(g) + CO_2(g) \rightleftharpoons H_2O(g) + CO(g)$

3 Give the units for K_p for each of the equilibria in question 2.

NOTE

If you are not sure about significant figures, see *Essential AS Chemistry for OCR*, Topic 16.2.

The effect of changing conditions on equilbria

Henri Louis Le Chatelier put forward his principle in 1888

In *Essential AS Chemistry for OCR*, Topic 15.2, we met **Le Chatelier's principle**, which we can use to predict the effect of changing temperature and pressure on the position of equilibrium. It states that when a system at equilibrium is disturbed, the equilibrium position moves in the direction that will reduce the disturbance.

So, if we increase the pressure on a gas phase equilibrium, the equilibrium moves towards the side with fewest molecules. If we decrease the temperature of a reaction, it will move in the exothermic direction (in which heat is given out). In this topic we look at what underlies this.

12.4.1 The effect of changing temperature on the equilibrium constant

Changing the temperature changes the value of the equilibrium constants, K_c or K_p. Whether K_c or K_p increases or decreases depends on whether the reaction is exothermic or endothermic. What happens is summarised in Table 12.1.

Table 12.1 The effect of changing temperature on equilibria

Type of reaction	Temperature change	Effect on K_c or K_p	Effect on products	Effect on reactants	Direction of change of equilibrium
endothermic	decrease	decrease	decrease	increase	moves left
endothermic	increase	increase	increase	decrease	moves right
exothermic	increase	decrease	decrease	increase	moves left
exothermic	decrease	increase	increase	decrease	moves right

> **HINT**
>
> When the value for $\Delta H^{\ominus}$ is given for a reversible reaction, it is taken to refer to the forward reaction i.e. left to right.

If the equilibrium constant K_c (or K_p) goes up, the equilibrium moves to the right (more product). If it goes down, the equilibrium moves left (less product).

The general rule is that

- for an exothermic reaction (ΔH is negative) increasing the temperature decreases the equilibrium constant
- for an endothermic reaction (ΔH is positive) increasing the temperature increases the equilibrium constant.

So for an exothermic reaction, increasing the temperature will move the equilibrium to the left and for an endothermic reaction, increasing the temperature will move the equilibrium to the right.

12.4.2 The effect of changing pressure on the equilibrium constant

Changing the pressure of a reaction at constant temperature will only affect reactions involving gases. The volume of solids and liquids is not affected by increased pressure. The volume of a gas, on the other hand, changes dramatically with pressure. So increasing the pressure (i.e. decreasing the volume) has a similar effect to increasing the concentration of a reaction in solution.

It is found by experiment that though the position of equilibrium may change with pressure and thus the amount of reactants and products:

the value of K_p is not affected by pressure.

The equilibrium position of a reaction between gases will only be changed by pressure when there is a different total number of moles on each side of the equation.

So in the reaction $\qquad$ $H_2(g) + I_2(g) \rightleftharpoons 2HI(g)$

$\qquad\qquad\qquad\qquad$ 2 moles $\qquad$ 2 moles

the equilibrium position *will not change* when the pressure is increased – the proportions of H_2, I_2 and HI will stay the same.

For this reaction $K_p = \dfrac{pHI(g)_{eqm}^2}{pH_2(g)_{eqm}\, pI_2(g)}$, and K_p itself is not affected by pressure.

Increasing the pressure increases the partial pressures of HI, H_2 and I_2, but by the same amount. The effect of these increases cancels out.

But in the equilibrium:

$$N_2O_4(g) \rightleftharpoons 2NO_2(g)$$

$\qquad$ dinitrogen tetraoxide $\qquad\qquad$ nitrogen dioxide

$\qquad\qquad$ 1 mole $\qquad\qquad\qquad\qquad$ 2 moles

increasing the pressure *will* change the composition of the equilibrium mixture. We can use Le Chatelier's principle to predict the effect of pressure. If we increase the pressure, the equilibrium will move in the direction that reduces the pressure i.e. to the left where there are fewer molecules of gas. The underlying reason for this is as follows.

$$K_p = \dfrac{pNO(g)_{eqm}^2}{pN_2O_4(g)_{eqm}}$$

If we increase the pressure, pNO_2 will increase and so will pN_2O_4. But, the pNO_2 term in the expression for K_p is squared and the increase will therefore have more effect than that of pN_2O_4. We know that K_p does not change with pressure, so the only way that K_p can remain the same is for some NO_2 to change into N_2O_4. This reduces pNO_2 and increases pN_2O_4. So, the equilibrium moves to the left (as Le Chatelier's principle predicts).

12.4.3 K_p and the position of equilibrium

The size of the equilibrium constant, K_p, can tell us about the composition of the equilibrium mixture. The equilibrium law expression is always of the general form $\dfrac{\text{products}}{\text{reactants}}$. So:

- If K_p is much greater than 1, products predominate over reactants. We usually say that the equilibrium is over to the right.
- If K_p is much less than 1, reactants predominate, and the equilibrium position is over to the left.

Reactions where the equilibrium constant is greater than 10^{10} are usually regarded as going to completion while those with an equilibrium constant of less than 10^{-10} are regarded as not taking place at all.

QUICK QUESTIONS

1 Predict the effect of increasing (i) the pressure and (ii) the temperature on the following reactions:

a $2SO_2 + O_2 \rightleftharpoons 2SO_3$ $\qquad\qquad\qquad$ $\Delta H = -197\ \text{kJ mol}^{-1}$

b $N_2O_4(g) \rightleftharpoons 2NO_2(g)$ $\qquad\qquad\quad$ $\Delta H = +58\ \text{kJ mol}^{-1}$

c $H_2(g) + CO_2(g) \rightleftharpoons H_2O(g) + CO(g)$ $\qquad$ $\Delta H = +40\ \text{kJ mol}^{-1}$

Acids, bases and buffers

Defining an acid

We met with the definition of an acid as proton donor in *Essential AS Chemistry for OCR*, Topic 15.4. This is the Lowry–Brønsted description of acidity (developed in 1923 by Thomas Lowry and Johannes Brønsted independently) and it is the most generally useful current theory.

> **An acid is a substance that can donate a proton (a H^+ ion) and a base is a substance that can accept a proton.**

13.1.1 Proton transfer

Hydrogen chloride gas and ammonia gas react together to form ammonium chloride – a white ionic solid:

$$HCl + NH_3 \rightarrow Cl^- + NH_4^+$$

| hydrogen chloride | ammonia | chloride ion | ammonium ion |

Here hydrogen chloride is acting as an acid by donating a proton to ammonia. Ammonia is acting as a base by accepting a proton. Acids and bases can only react in pairs – one acid and one base.

Conjugate acid–base pairs

The Cl^- ion left after HCl has donated a proton is itself able to accept a proton (to go back to HCl) and is therefore a base. It is called the **conjugate base** of HCl. In the same way, the NH_4^+ ion is an acid as it can donate a proton to return to being NH_3. It is the **conjugate acid** of ammonia. So we have two conjugate acid–base pairs. These can be written either way round:

$$HCl \quad \text{and} \quad Cl^- \quad \text{or} \quad HCl \quad \text{and} \quad Cl^-$$

| acid 1 | conjugate base 1 | conjugate acid 1 | base 1 |

and

$$NH_4^+ \quad \text{and} \quad NH_3 \quad \text{or} \quad NH_4^+ \quad \text{and} \quad NH_3$$

| conjugate acid 2 | base 2 | acid 2 | conjugate base 2 |

So, we could think of the reaction in these terms:

$$HCl + NH_3 \rightarrow Cl^- + NH_4^+$$

| acid 1 | base 2 | conjugate base 1 | conjugate acid 2 |

HCl is a much better acid (proton donor) than NH_4^+ as it is able to give away its proton and force NH_3 to accept it.

13.1.2 Water as an acid and a base

HCl can donate a proton to water, so that water acts as a base:

$$HCl + H_2O \rightarrow H_3O^+ + Cl^-$$

These are the species involved:

$$HCl \quad \text{and} \quad Cl^- \quad \text{(or } HCl \quad \text{and} \quad Cl^-\text{)}$$

| acid | conjugate base | conjugate acid | base |

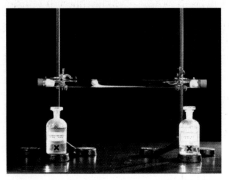

Fig 13.1 *The white ring of ammonium chloride is formed when hydrogen chloride (left) and ammonia (right) react*

and

$$H_3O^+ \quad \text{and} \quad H_2O \qquad \text{(or } H_3O^+ \text{ and } H_2O)$$

$$\text{acid} \qquad \text{conjugate} \qquad \text{conjugate} \qquad \text{base}$$
$$\text{base} \qquad \text{acid}$$

H_3O^+ is called the **oxonium ion** but the names hydronium ion and hydroxonium ion are also used.

Water may also act as an acid. For example:

$$H_2O + NH_3 \rightarrow OH^- + NH_4^+$$

Here OH^- is the conjugate base and H_2O the acid.

13.1.3 Competition for protons

Acid and base reactions can be thought of as competitions for protons, the better base 'winning' the proton.

$$\text{So in} \qquad HCl \quad + \quad NH_3 \quad \rightarrow \quad Cl^- \quad + \quad NH_4^+$$

$$\text{acid 1} \qquad \text{base 2} \qquad \text{conjugate} \qquad \text{conjugate}$$
$$\text{base 1} \qquad \text{acid 2}$$

ammonia must be a better base than the chloride ion, as it 'wins' the proton.

13.1.4 The proton in aqueous solution

It is important to realise that a H^+ ion is just a proton. The hydrogen atom has only one electron and if this is lost all that remains is a proton (the hydrogen nucleus). This is about 10^{-15} m in diameter, compared to 10^{-10} m or more for any other chemical entity. This extremely small size and consequent intense electric field cause it to have unusual properties compared with other positive ions. It is never found isolated. In aqueous solutions it is always bonded to at least one water molecule to form the ion H_3O^+. For simplicity, we shall represent a proton in an aqueous solution by $H^+(aq)$ rather than $H_3O^+(aq)$.

Since a H^+ ion has no electrons of its own, it can only form a bond with another species that has a lone pair of electrons.

13.1.5 The ionisation of water

Water is slightly ionised:

$$H_2O(l) \rightleftharpoons H^+(aq) + OH^-(aq)$$

or this may be written:

$$H_2O(l) + H_2O(l) \rightleftharpoons H_3O^+(aq) + OH^-(aq)$$

to emphasise that this is an acid–base reaction in which one water molecules donates a proton to another.

This equilibrium is established in water and all aqueous solutions:

$$H_2O(l) \rightleftharpoons H^+(aq) + OH^-(aq)$$

and we can write an equilibrium expression, see Topic 12.1:

$$K_c = \frac{[H^+(aq)]_{eqm}[OH^-(aq)]_{eqm}}{[H_2O(l)]_{eqm}}$$

The concentration of water, $[H_2O(l)]$, is constant and is incorporated into a modified equilibrium constant K_w, where $K_w = K_c \times [H_2O(l)]$.

So, $K_w = \mathbf{[H^+(aq)]_{eqm} [OH^-(aq)]_{eqm}}$

> **HINT**
>
> The important point to remember at this stage is that the product of $[H^+(aq)]_{eqm}$ and $[OH^-(aq)]_{eqm}$ is constant so that if one goes up, the other must go down proportionately.

K_w is called the **ionic product** of water and at 298 K it is equal to 10^{-14} mol^2 dm^{-6}. Each H_2O that dissociates (splits up) gives rise to one H^+ and one OH^- so, in pure water, at 298 K:

$$[OH^-(aq)] = [H^+(aq)]$$
$$\text{So } 10^{-14} = [H^+(aq)]^2$$
$$[H^+(aq)] = 10^{-7} mol\, dm^{-2} = [OH^-(aq)]$$

> **DID YOU KNOW?**
>
> $[H^+(aq)] = 10^{-7}$ mol dm^{-3} means that only 2 in every 1 000 000 000 water molecules is split up into hydrogen and hydroxide ions. To put this in context, 1 000 000 000 is the number of letters in about 3000 copies of this book!

QUICK QUESTIONS

1 a Hydrogen bromide, HBr, is acidic. What is its conjugate base?

b OH^- is a base. What is its conjugate acid?

2 Identify the conjugate acid/base pairs in
$$HNO_3 + OH^- \rightarrow NO_3^- + H_2O$$

3 Magnesium oxide is describes as basic, whereas sodium hydroxide is an alkali. Explain the difference?

4 In an acidic solution, $[H^+]$ is 10^{-4} mol dm^{-3}. What is $[OH^-(aq)]$?

5 What species are formed when the following bases accept a proton?

a OH^-

b NH_3

c H_2O

d Cl^-

The pH scale

The acidity of a solution depends on the concentration of $H^+(aq)$ and is measured on the pH scale.

> ## pH is defined as $-\log_{10}[H^+(aq)]$

Remember that square brackets, [ ], mean the concentration in $mol\,dm^{-3}$.

Although this expression is more complicated than simply stating the concentration of $H^+(aq)$, it does away with awkward numbers like 10^{-13}, etc, which occur because the concentration of $H^+(aq)$ in most aqueous solutions is so small. The minus sign makes almost all pH values positive (because the logs of numbers less than 1 are negative).

On the pH scale:

- the *smaller* the pH, the *greater* the concentration of $H^+(aq)$.
- a difference of *one* pH number means a *tenfold* difference in acidity, so that pH2 is ten times as acidic as pH3.

Remember $K_w = [H^+(aq)]_{eqm}[OH^-(aq)]_{eqm} = 10^{-14}\,mol^2\,dm^{-6}$. This means that in neutral aqueous solutions:

$$[H^+(aq)] = [OH^-(aq)] = 10^{-7}\,mol\,dm^{-3}$$
$$pH = -\log_{10}[H^+(aq)] = -\log_{10}[10^{-7}] = 7$$

so the pH is 7.

pH measures alkalinity as well as acidity, because as $[H^+(aq)]$ goes up, $[OH^-(aq)]$ goes down. If a solution contains more $H^+(aq)$ than $OH^-(aq)$, its pH will be less than 7 and we call it acidic. If a solution contains more $OH^-(aq)$ than $H^+(aq)$, its pH will be greater than 7 and we call it alkaline, see Figure 13.1.

13.2.1 Working with the pH scale

Finding $[H^+(aq)]$ from pH

We can work out the concentration of hydrogen ions, $[H^+]$, in an aqueous solution if we know the pH. It is the anti-log of the pH value. For example, an acid has a pH of 3.

$$pH = -\log_{10}[H^+(aq)]$$
$$3 = -\log_{10}[H^+(aq)]$$
$$-3 = \log_{10}[H^+(aq)]$$

Take the antilog of both sides:

$$[H^+(aq)] = 10^{-3}\,mol\,dm^{-3}$$

With bases, we need two steps. Suppose the pH of a solution is 10

$$pH = -\log_{10}[H^+(aq)]$$
$$10 = -\log_{10}[H^+(aq)]$$
$$-10 = \log_{10}[H^+(aq)]$$

Take the antilog of both sides:

So, $\qquad [H^+(aq)] = 10^{-10}$

We know $\qquad [H^+(aq)]\,[OH^-(aq)] = 10^{-14}\,mol^2\,dm^{-6}$

Substituting our value for $[H^+(aq)] = 10^{-10}$ into the equation:

$$[10^{-10}]\,[OH^-(aq)] = 10^{-14}\,mol^2\,dm^{-6}$$

So, $\qquad [OH^-(aq)] = 10^{-14}/10^{-10} = 10^{-4}\,mol\,dm^{-3}$

> **DID YOU KNOW?**
>
> Distilled water is usually slightly acidic because it has carbon dioxide dissolved in it.

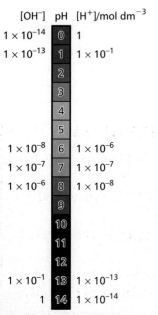

Fig 13.1 *The pH scale. What is the concentration of stomach acid (hydrochloric acid) which has a pH of 2?*

Calculating the pH from a concentration

We can calculate the pH if we know the concentration of acid or alkali.

The basic method is to work out $[H^+(aq)]$ (even if the solution is alkaline) and then use $pH = -\log_{10}[H^+(aq)]$.

The pH of strong acid solutions

HCl dissociates completely in water to $H^+(aq)$ ions and $Cl^-(aq)$ ions, i.e. the reaction:

$$HCl(aq) \rightarrow H^+(aq) + Cl^-(aq)$$

goes to completion. Acids that dissociate completely like this are called strong acids, see Topic 13.3.

So in 1 mol dm^{-3} HCl,

$$[H^+(aq)] = 1 \text{ mol dm}^{-3}$$
$$\log[H^+(aq)] = \log 1 = 0$$
$$-\log[H^+(aq)] = 0$$

so the pH of 1 mol dm^{-3} HCl = 0.

In a 0.1 mol dm^{-3} solution of HCl,

$$[H^+(aq)] = 0.1 \text{ mol dm}^{-3}$$
$$\log[H^+(aq)] = \log 0.1 = -1$$
$$-\log[H^+(aq)] = 1$$

So the pH of 0.1 mol dm^{-3} HCl = 1.

The pH of alkaline solutions

In alkaline solutions, it takes two steps to calculate $[H^+(aq)]$.

To find the $[H^+(aq)]$ of an alkaline solution:

- calculate $[OH^-(aq)]$,
- then use: $[H^+(aq)][OH^-(aq)] = 10^{-14} \text{ mol}^2 \text{ dm}^{-6}$ to calculate $[H^+(aq)]$.

The pH can then be calculated.

For example to find the pH of 1 mol dm^{-3} sodium hydroxide solution:

Sodium hydroxide is fully dissociated in aqueous solution – we call this a strong alkali, see Topic 13.3.

$$NaOH(aq) \rightarrow Na^+(aq) + OH^-(aq)$$

so $\quad [OH^-(aq)] = 1 \text{ mol dm}^{-3}$

but $\quad [OH^-(aq)][H^+(aq)] = 10^{-14} \text{ mol dm}^{-3}$
$$1 \times [H^+(aq)] = 10^{-14} \text{ mol dm}^{-3}$$
$$\log[H^+(aq)] = -14$$
$$pH = 14.$$

In a 0.1 mol dm^{-3} sodium hydroxide solution

$$[OH^-(aq)] = 10^{-1} \text{ mol dm}^{-3}$$
$$[OH^-(aq)][H^+(aq)] = 10^{-14} \text{ mol dm}^{-3}$$
$$[H^+(aq)] \times 10^{-1} = 10^{-14} \text{ mol dm}^{-3}$$
$$[H^+(aq)] = 10^{-13} \text{ mol dm}^{-3}$$
$$\log[H^+(aq)] = -13$$
$$pH = 13.$$

QUICK QUESTIONS

1 What is the pH of a solution in which $[H^+]$ is $10^{-2} \text{ mol dm}^{-3}$?

2 What is $[H^+]$ in a solution of $pH = 6.0$?

3 What is $[OH^-]$ in a solution of $pH = 9.0$?

4 Calculate the pH of a $0.020 \text{ mol dm}^{-3}$ solution of HCl.

5 Calculate the pH of 0.20 mol dm^{-3} sodium hydroxide.

13.3 Finding the pH of weak acids and bases

NOTE

Although in the gas phase hydrogen chloride, HCl, is a covalent molecule, a solution of it in water is wholly ionic – ($H^+(aq)$ + $Cl^-(aq)$). We can assume that there are no molecules remaining, so hydrochloric acid is a strong acid.

HINT

The strength of an acid and its concentration are completely independent, so use the two different words carefully.

In Topic 13.2 we found the pH of acids such as hydrochloric acid which, when dissolved in water, dissociate completely into ions. Acids that completely dissociate into ions in aqueous solutions are called **strong acids**. The word strong refers *only* to the extent of dissociation and *not in any way* to the concentration. So it is perfectly possible to have a very dilute solution of a strong acid.

The same arguments apply to bases. Strong bases are completely dissociated into ions in aqueous solutions. For example sodium hydroxide is a strong base:

$$NaOH(aq) \rightarrow Na^+(aq) + OH^-(aq)$$

In this topic we look at weak acids and bases and see how to find their pHs.

13.3.1 Weak acids and bases

Many acids and bases are not fully dissociated when dissolved in water. Ethanoic acid (the acid in vinegar, also known as acetic acid) is a typical example. In a $1\ mol\ dm^{-3}$ solution of ethanoic acid, only about 4 in every thousand ethanoic acid molecules are dissociated into ions (so the degree of dissociation is 4/1000); the rest remain dissolved as wholly covalently bonded molecules. In fact an equilibrium is set up:

	$CH_3CO_2H(aq)$ $\rightleftharpoons$	$H^+(aq)$ +	$CH_3CO_2^-(aq)$
	ethanoic acid	hydrogen ions	ethanoate ions
Before dissociation	1000	0	0
At equilibrium	996	4	4

Acids like this are called **weak acids**. Again note that weak refers *only* to the degree of dissociation. In a $5\ mol\ dm^{-3}$ solution, ethanoic acid is still a weak acid, while in a $10^{-4}\ mol\ dm^{-3}$ solution, hydrochloric acid is still a strong acid.

Ammonium hydroxide (NH_4OH) is a weak base and the equilibrium lies well to the left:

$$NH_4OH(aq) \rightleftharpoons NH_4^+(aq) + OH^-(aq)$$

13.3.2 The dissociation of weak acids and bases

Weak acids

Imagine a weak acid HA which dissociates:

$$HA(aq) \rightleftharpoons H^+(aq) + A^-(aq)$$

The equilibrium constant is given by:

$$K_c = \frac{[H^+(aq)]_{eqm}[A^-(aq)]_{eqm}}{[HA(aq)]_{eqm}}$$

This could equally well be written:

$$HA(aq) + H_2O(l) \rightleftharpoons H_3O^+(aq) + A^-(aq)$$

$$K'_c = \frac{[H_3O^+(aq)]_{eqm}[A^-(aq)]_{eqm}}{[HA(aq)]_{eqm}[H_2O(l)]_{eqm}}$$

Since the concentration of $H_2O(l)$ is effectively constant, it is incorporated into the value of K_c (just as K_w for the dissociation of water incorporates $[H_2O(l)]$):

$$K_c = \frac{[H^+(aq)]_{eqm}[A^-(aq)]_{eqm}}{[HA(aq)]_{eqm}}$$

where $K_c = K'_c[H_2O(l)]$

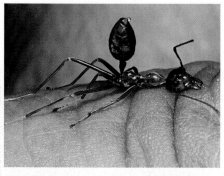
An ant spraying formic acid. Formic acid (methanoic acid) is quite concentrated when used as a weapon by the stinging ant and, although it is a weak acid, being sprayed with it can be a painful experience

For a weak acid, this is usually given the symbol K_a and called the acid dissociation constant.

$$K_a = \frac{[H^+(aq)]_{eqm}[A^-(aq)]_{eqm}}{[HA(aq)]_{eqm}}$$

The larger the value of K_a, the further the equilibrium is to the right, the more the acid is dissociated and the stronger it is. Acid dissociation constants for some acids are given in Table 13.1.

Table 13.1 Values of K_a for some weak acids

Acid	K_a/mol dm^{-3}
chloroethanoic	1.3×10^{-3}
benzoic	6.3×10^{-5}
ethanoic	1.7×10^{-5}
hydrocyanic	4.9×10^{-10}

13.3.3 Calculating the pH of solutions weak acids

In Topic 13.2 we calculated the pH of solutions of strong acids, by assuming that they are fully dissociated. For example, in a 1 mol dm^{-3} solution of nitric acid, $[H^+] = 1$ mol dm^{-3}. In weak acids this is no longer true and we must use the equilibrium law to calculate $[H^+]$.

Calculating the pH of 1 mol dm^{-3} ethanoic acid:
Using the same method as in Topic 12.2 for equilibrium calculations, the concentrations in mol dm^{-3} are:

$$CH_3CO_2H(aq) \rightleftharpoons CH_3CO_2^-(aq) + H^+(aq)$$

Before dissociation: 1   0   0

At equilibrium: $1 - [CH_3CO_2^-(aq)]$ $[CH_3CO_2^-(aq)]$ $[H^+(aq)]$

$$K_a = \frac{[CH_3CO_2^-(aq)]_{eqm}[H^+(aq)]_{eqm}}{[CH_3CO_2H(aq)]_{eqm}}$$

But as each CH_3CO_2H molecule that dissociates produces one $CH_3CO_2^-$ ion and one H^+ ion:

$$[CH_3CO_2^-(aq)]_{eqm} = [H^+(aq)]_{eqm}$$

$$K_a = \frac{[H^+(aq)]_{eqm}^2}{1 - [H^+(aq)]_{eqm}}$$

Since the degree of dissociation of ethanoic acid is so small (it is a weak acid), $[H^+(aq)]_{eqm}$ is very small and to a good approximation, $1 - [H^+(aq)]_{eqm} \approx 1$.

so $K_a = \frac{[H^+(aq)]_{eqm}^2}{1}$

From Table 13.1, $K_a = 1.7 \times 10^{-5}$ mol dm^{-3}

so $1.7 \times 10^{-5} = [H^+(aq)]_{eqm}^2$

$$[H^+(aq)]_{eqm} = \sqrt{1.7 \times 10^{-5}}$$

$$[H^+(aq)]_{eqm} = 4.12 \times 10^{-3} \text{ mol dm}^{-3}$$

Taking logs: $\log[H^+(aq)]_{eqm} = -2.384$

so   pH = 2.384.

Calculating the pH of 0.1 mol dm^{-3} ethanoic acid:

Using the same method, we get:

$$K_a = \frac{[H^+(aq)]_{eqm}^2}{0.1 - [H^+(aq)]_{eqm}}$$

Again, $0.1 - [H^+(aq)]_{eqm} \approx 0.1$

so: $1.7 \times 10^{-5} = \frac{[H^+(aq)]_{eqm}^2}{0.1}$

$$1.7 \times 10^{-6} = [H^+(aq)]_{eqm}^2$$

$$[H^+(aq)]_{eqm} = 1.3 \times 10^{-3} \text{ mol dm}^{-3}$$

$$pH = 2.88$$

13.3.4 pK_a

For a weak acid we use pK_a. This is defined as:

$$pK_a = -\log_{10}K_a$$

Think of 'p' as meaning '$-\log_{10}$ of'. pK_a can be useful in calculations, see Table 13.2. It gives a measure of how strong a weak acid is – the smaller the value of pK_a, the stronger the acid.

Table 13.2 Values of K_a and pK_a for some weak acids

Acid	K_a/mol dm^{-3}	pK_a
chloroethanoic	1.3×10^{-3}	2.88
benzoic	6.3×10^{-5}	4.20
ethanoic	1.7×10^{-5}	4.77
hydrocyanic	4.9×10^{-10}	9.31

QUICK QUESTIONS

1 Which is the strongest acid in Table 13.2.?

2 What is the concentration of undissociated HCl molecules in a 1 mol dm^{-3} solution of hydrochloric acid?

3 What is the pH of the following solutions:
   a 0.1 mol dm^{-3} chloroethanoic acid;
   b 0.01 mol dm^{-3} benzoic acid?

4 What can you say about the concentration of H$^+$ ions and the concentration of ethanoate ions in all solutions of ethanoic acid?

Acid–base titrations

A titration is used to find the concentration of a solution by gradually adding to it a second solution of known concentration with which it reacts. To use a titration, we must know the equation for the reaction.

13.4.1 pH changes during acid–base titrations

In an acid–base titration, a solution of a base (an alkali) of known concentration is added from a burette to a measured amount of a solution of an acid until an indicator shows that the acid has been neutralised. We can then work out the concentration of the acid from the volume of alkali used. We can also follow a neutralisation reaction by measuring the pH with a pH meter, see Figure 13.2.

An acid–base titration, to find the concentration of a base. A **pipette** is used to deliver an accurately measured volume of acid of unknown concentration into the flask. The alkali of known concentration is in the **burette**

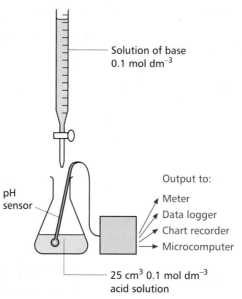

Solution of base
0.1 mol dm^{-3}

pH sensor

Output to:
- Meter
- Data logger
- Chart recorder
- Microcomputer

25 cm^3 0.1 mol dm^{-3} acid solution

Fig 13.2 *Apparatus to investigate pH changes during a titration*

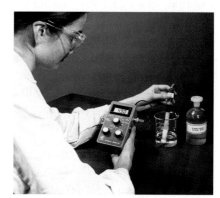

A pH meter

13.4.2 Titration curves

Figures 13.3 (a), (b), (c) and (d) show the results obtained for four cases. Notice that in these cases the base is added from the burette and the acid has been accurately measured into a flask. The shape of each titration curve is typical for the type of acid–base titration.

The first thing to notice about these curves is that the pH does *not* change in a regular manner as the acid is added. Each curve has almost horizontal sections where a lot of base can be added without changing the pH much. There is also a very steep portion of each curve (except weak acid/weak base) where a single drop of base changes the pH by several units.

In a titration, the **equivalence point** is the point at which exactly the same number of moles of hydroxide ions have been added as there are moles of hydrogen ions. In each of the titrations in Figure 13.3 the equivalence point is reached after 25 cm^3 of base has been added. However, the pH at the equivalence point is not always exactly 7.

Notice that in each case, except the weak acid–weak base titration, there is a large and rapid change of pH at the equivalence point (i.e. the curve is almost vertical) even though this is may not be centred on pH 7. This is relevant to the choice of indicator for a particular titration, see Topic 13.5.

QUICK QUESTIONS

Quick questions for this topic will be found at the end of Topic 13.5.

(a)
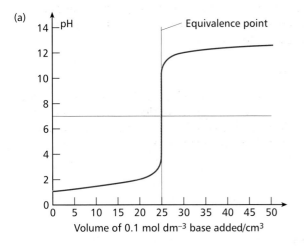

(c)
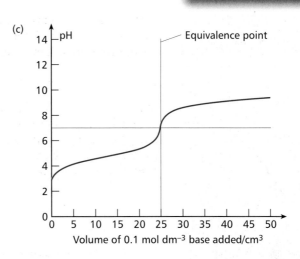

(b)
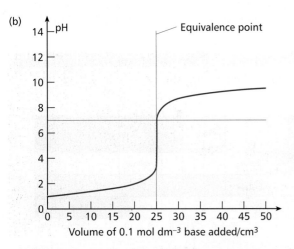

(d)
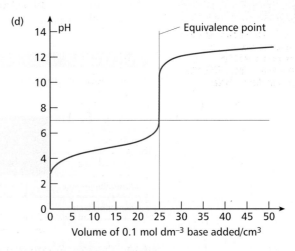

Fig 13.3 *Graphs of pH changes for titrations of different acids and bases. In each case, we start with 25 cm³ of the acid in the flask and add the base from a burette. Titration of (a) strong acid and strong base, (b) strong acid and weak base, (c) weak acid and weak base, (d) weak acid and strong base*

An acid–base titration uses an **indicator** to find the concentration of a solution of an acid or alkali. The **end point** is the volume of alkali or acid added when the indicator just changes colour.

A suitable indicator for a particular titration needs the following properties:

- The colour change must be sharp rather than gradual at the end point, i.e. no more than 1 drop of acid (or alkali) is needed to give a complete colour change. An indicator that changes colour gradually over several cubic centimetres would be of no use.
- The end point of the titration given by the indicator must be the same as the equivalence point. Otherwise the titration will give us the wrong answer.
- The indicator should give a distinct colour change. For example, the colourless to pink change of phenolphthalein is easier to see than the red to yellow of methyl orange.

Some common indicators are given in the box *Indicators*. Notice that the colour change of most indicators takes place over a pH range of around 2 units, centred around different pHs. For this reason not all indicators are suitable for all titrations.

INDICATORS

Some common indicators are shown below with their approximate colour changes.

Universal indicator															
Methyl orange		Red		Change				Yellow							
Bromophenol blue			Yellow	Change			Blue								
Phenolphthalein				Colourless				Change		Red					
	0	1	2	3	4	5	6	7	8	9	10	11	12	13	14	pH

very acidic   neutral   very alkaline

Universal indicator is a mixture of indicators that change colour at different pHs.

The following examples compare the suitability of two common indicators, phenolphthalein and methyl orange, for examples of four different types of acid–base titrations. In each case the base is being added to the acid.

1. Strong acid–strong base – for example hydrochloric acid and sodium hydroxide

Figure 13.4 is the graph of pH against volume of base added. The pH ranges, over which the two indicators change colour, are shown. To fulfil the first two criteria above, the indicator must change within the vertical portion of the pH curve. Here either indicator would be suitable but phenolphthalein is usually preferred because of its more easily seen colour change.

2. Weak acid–strong base titration – for example ethanoic acid and sodium hydroxide

Methyl orange is not suitable, see Figure 13.5. It does not change in the vertical portion of the curve. Phenolphthalein will change sharply at exactly 25 cm^3, the equivalence point, and would therefore be a good choice.

Strong acid–weak base titration – for example hydrochloric acid and ammonia

Here methyl orange will change sharply at the equivalence point but phenolphthalein would be of no use, see Figure 13.6.

Weak acid–weak base – for example ethanoic acid and ammonia

Here neither indicator is suitable, see Figure 13.7. In fact no indicator could be suitable as an indicator requires a vertical portion of the curve over two pH units at the equivalence point to give a sharp change.

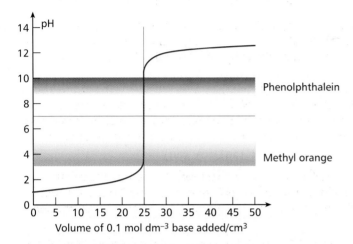

Fig 13.4 *Titration of a strong acid–strong base, adding 0.1 mol dm⁻³ NaOH(aq) to 25 cm³ of 0.1 mol dm⁻³ HCl(aq)*

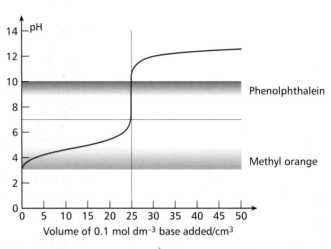

Fig 13.5 *Titration of a weak acid–strong base, adding 0.1 mol dm⁻³ NaOH(aq) to 25 cm³ of 0.1 mol dm⁻³ CH₃CO₂H(aq)*

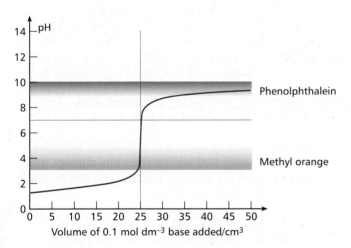

Fig 13.6 *Titration of a strong acid–weak base, adding 0.1 mol dm⁻³ NH₃(aq) to 25 cm³ of 0.1 mol dm⁻³ HCl(aq)*

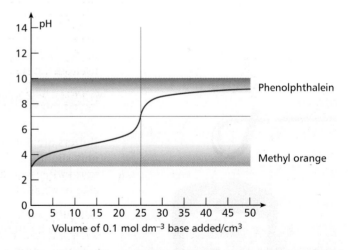

Fig 13.7 *Titration of a weak acid–weak base, adding 0.1 mol dm⁻³ NH₃(aq) to 25 cm³ of 0.1 mol dm⁻³ CH₃CO₂H(aq)*

QUICK QUESTIONS

1 Why would universal indicator not be suitable for *any* titration?

2 The indicator bromocresol purple changes colour between pH 5.2 and 6.8. For which of the following titration types would it be suitable:

  a weak acid–weak base   **c** weak acid–strong base

  b strong acid–weak base   **d** strong acid–strong base?

Buffer solutions

Buffers are solutions that can resist changes of acidity or alkalinity. When small amounts of acid or alkali are added to them, their pH remains almost constant.

13.6.1 How buffers work

Some buffers are based on weak acids. They work because the dissociation of a weak acid is an equilibrium reaction.

Imagine a weak acid HA. It will dissociate in solution:

$$HA(aq) \rightleftharpoons H^+(aq) + A^-(aq)$$

As it is a weak acid $[H^+(aq)] = [A^-(aq)]$ and is small.

If a little alkali is added, the OH^- ions will react with H^+ ions to remove them as water molecules:

$$HA(aq) \rightleftharpoons H^+(aq) + A^-(aq)$$
$$\downarrow OH^-(aq)$$
$$H_2O$$

This disturbs the equilibrium and (as predicted by Le Chatelier's principle, *Essential AS Chemistry for OCR*, Topic 15.2) more HA will dissociate to restore the situation, so the pH tends to remain the same. If more H^+ is added, again the equilibrium shifts, this time to the left, H^+ ions combining with A^- ions to produce undissociated HA.

However, since $[A^-]$ is small, the supply of A^- soon runs out and there is no A^- left to 'mop up' the added H^+. So we have half a buffer!

However, we can add to the solution a supply of extra A^- by adding a soluble salt of HA such as Na^+A^-. This increases the supply of A^- so that more H^+ can be used up. So there is a way in which both added H^+ and OH^- can be removed. So:

> **A buffer may be made from a mixture of a weak acid and a soluble salt of that acid.**

Buffers are not perfect. The addition of acid or alkali will still change the pH, but only slightly and by far less than the change that adding the same amount to a non-buffer would cause. It is also possible to 'saturate' a buffer – to add so much acid or alkali that all of the available H^+ or A^- is used up.

The function of the weak acid component of a buffer is to act as a source of H^+ ions which can mop up any added OH^-:

$$HA + OH^- \rightarrow A^- + H_2O$$

The function of the salt component of a buffer is to mop up any added H^+ ions:

$$A^- + H^+ \rightarrow HA$$

13.6.2 A different type of buffer

A mixture of ammonium chloride ($NH_4^+Cl^-$) and ammonia also acts as a buffer. In this case:

- the ammonium ion, NH_4^+, acts as the weak acid, dissociating to provide H^+ ions to mop up added OH^-:

$$NH_4^+(aq) \rightleftharpoons NH_3(aq) + H^+(aq)$$
$$\downarrow OH^-(aq)$$
$$H_2O(1)$$

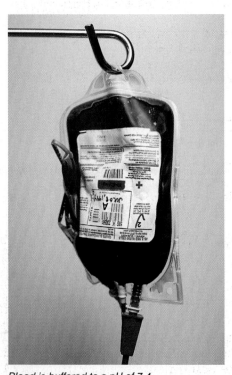

Blood is buffered to a pH of 7.4

- the ammonia mops up added H^+:

$$NH_3(aq) + H^+(aq) \rightarrow NH_4^+(aq)$$

Another example of a system involving a buffer is blood, whose pH is maintained at approximately 7.4. A change of as little as 0.5 of a pH unit may well be fatal.

Blood is buffered to a pH of 7.4 by a number of mechanisms. The most important is:

$$H^+(aq) + HCO_3^-(aq) \rightleftharpoons CO_2(aq) + H_2O(l)$$

Addition of extra H^+ ions moves this equilibrium to the right, thus removing the added H^+. Addition of extra OH^- ions removes H^+ by reacting to form water. The equilibrium above moves to the left releasing more H^+ ions. The same equilibrium helps to buffer the acidity of soils.

There are many examples of buffers in everyday products, such as detergents and shampoos. If either of these products become too acidic or too alkaline, they could damage fabric (or skin and hair). Typically, the buffers they use are based on phosphoric acid and sodium phosphate.

13.6.3 Calculations on buffers

Different buffers can be made which will maintain different pHs. When a weak acid dissociates:

$$HA(aq) \rightleftharpoons H^+(aq) + A^-(aq)$$

we can write the expression:

$$K_a = \frac{[H^+(aq)]_{eqm}[A^-(aq)]_{eqm}}{[HA(aq)]_{eqm}}$$

This can be rearranged (see the box *Deriving the Henderson equation*) to give the Henderson equation, which is more useful for calculating the pH of a buffer.

$$pH = pK_a - \log\left(\frac{[HA(aq)]}{[A^-(aq)]}\right)$$

Calculating the pH of a buffer

A buffer consists of 0.1 mol dm^{-3} ethanoic acid and 0.1 mol dm^{-3} sodium ethanoate. pK_a for ethanoic acid is 4.8. What is its pH?

Using the Henderson equation:

$$pH \text{ of buffer} = 4.8 - \log\frac{(0.1)}{(0.1)}$$
$$= 4.8 - \log 1$$
$$pH = 4.8 - 0 = 4.8$$

Changing the concentration of HA or A^- will affect the pH of the buffer. If we use 0.2 mol dm^{-3} ethanoic acid and 0.1 mol dm^{-3} sodium ethanoate, the pH will be given by:

$$pH = 4.8 - \log\frac{(0.2)}{(0.1)}$$
$$= 4.8 - \log 2$$
$$pH = 4.8 - 0.3 = 4.5$$

Shampoos contain buffers to keep their pHs slightly alkaline

DERIVING THE HENDERSON EQUATION

You will not be examined on this derivation.

$$K_a = \frac{[H^+(aq)]_{eqm}[A^-(aq)]_{eqm}}{[HA(aq)]_{eqm}}$$

This may be written:

$$K_a = [H^+(aq)] \times \frac{[A^-(aq)]}{[HA(aq)]}$$

Taking logs and remembering that *multiplication* of numbers is achieved by *adding* logs:

$$\log K_a = \log[H^+(aq)] + \log\left(\frac{[A^-(aq)]}{[HA(aq)]}\right)$$

Multiplying both sides by -1:

$$-\log K_a = -\log[H^+(aq)] - \log\left(\frac{[A^-(aq)]}{[HA(aq)]}\right)$$

$-\log[H^+(aq)]$ is the pH of the solution and $-\log K_a$ can be called pK_a.

So
$$pK_a = pH - \log\left(\frac{[A^-(aq)]}{[HA(aq)]}\right)$$

or
$$pH = pK_a + \log\left(\frac{[A^-(aq)]}{[HA(aq)]}\right)$$

or
$$pH = pK_a - \log\left(\frac{[HA(aq)]}{[A^-(aq)]}\right)$$

This is called the Henderson equation.

QUICK QUESTIONS

1 Find the pH of the following buffers:

   a using [ethanoic acid] = 0.1 mol dm^{-3}, [sodium ethanoate] = 0.2 mol dm^{-3} (pK_a of ethanoic acid = 4.8).

   b using [benzoic acid] = 0.1 mol dm^{-3}, [sodium ethanoate] = 0.1 mol dm^{-3} (pK_a of benzoic acid = 4.2).

OCR Exam Questions
'Unifying concepts in Chemistry'

1 Nitrogen oxides such as nitrogen monoxide, NO, and nitrogen dioxide, NO_2, are formed unintentionally by man and cause considerable harm to the environment.

  a The oxidation of nitrogen monoxide in car exhausts may involve the following reaction.

  $$NO(g) + CO(g) + O_2(g) \rightarrow NO_2(g) + CO_2(g)$$

  This reaction was investigated in a series of experiments. The results are shown in the table below.

Experiment	[NO(g)] /mol dm^{-3}	[CO(g)] /mol dm^{-3}	[O$_2$(g)] /mol dm^{-3}	initial rate /mol dm^{-3} s^{-1}
1	1.00×10^{-3}	1.00×10^{-3}	1.00×10^{-1}	0.44×10^{-3}
2	2.00×10^{-3}	1.00×10^{-3}	1.00×10^{-1}	1.76×10^{-3}
3	2.00×10^{-3}	2.00×10^{-3}	1.00×10^{-1}	1.76×10^{-3}
4	2.00×10^{-3}	2.00×10^{-3}	4.00×10^{-1}	7.04×10^{-3}

  (i) For each reactant, deduce the order of reaction. Show your reasoning.
  (ii) Deduce the rate equation and calculate the rate constant for this reaction.
  (iii) Suggest, with a reason, what would happen to the value of the rate constant, k, as the car's exhaust gets hotter. [11]

  b State **two** environmental consequences of nitrogen oxides. [2]

  c Not all nitrogen compounds are harmful: some, such as nitrogen fertilisers, are beneficial to man.

  A nitrogen fertiliser, **D**, was analysed in the laboratory and was shown to have the composition by mass Na, 27.1%; N, 16.5%; O, 56.4%. On heating, 3.40 g of **D** was broken down into sodium nitrite, $NaNO_2$, and oxygen gas.

  Showing your working, suggest an identity for the fertiliser, **D**, and calculate the volume of oxygen that was formed.
  [Under the experimental conditions, 1 mole of gas molecules occupy 24 dm^3.] [4]

  [Total: 17]

  OCR, Specimen 2000

2 This question refers to different aspects of acid/base chemistry:

  a Hydrochloric acid HCl is classed as a **strong** acid but it can have both **concentrated** and **dilute** solutions. Explain why this is so. [3]

  b Sodium phosphate, Na_3PO_4, a water-softening agent, can be prepared in the laboratory by neutralising phosphoric acid.

A student prepared this compound in the laboratory from 20.0 cm^3 of 0.100 mol dm^{-3} phosphoric acid and 0.250 mol dm^{-3} sodium hydroxide.

$$H_3PO_4(aq) + 3NaOH(aq) \rightarrow Na_3PO_4(aq) + 3H_2O(l)$$

  (i) Deduce the oxidation state of phosphorus in sodium phosphate, Na_3PO_4.
  (ii) Calculate the volume of NaOH(aq) that the student would need to use to just neutralise the phosphoric acid using the quantities above. [4]

  c Calculate the pH of the NaOH(aq) used in **b** ($K_w = 1.00 \times 10^{-14}$ mol^2 dm^{-6}.) [4]

  [Total: 11]

  OCR, Specimen 2000

3 Bromine can be formed by the oxidation of bromide ions. This question compares the rates of two reactions that produce bromine.

  a Bromine is formed by the oxidation of bromide ions with acidified bromate(V) ions.

  $$5Br^-(aq) + 6H^+(aq) + BrO_3^-(aq) \rightarrow 3Br_2(aq) + 3H_2O(l)$$

  This reaction was carried out several times using different concentrations of the three reactants. The initial rate of each experimental run was calculated and the results are shown below. In each case, initial concentrations are shown.

Experiment	[Br$^-$(aq)] /mol dm^{-3}	[H$^+$(aq)] /mol dm^{-3}	[BrO$_3^-$(aq)] /mol dm^{-3}	initial rate /10^{-3} mol dm^{-3} s^{-1}
1	0.10	0.10	0.10	1.2
2	0.10	0.10	0.20	2.4
3	0.30	0.10	0.10	3.6
4	0.10	0.20	0.20	9.6

  (i) For each reactant, deduce the order of reaction. Show your reasoning. [6]
  (ii) Deduce the rate equation. [1]
  (iii) Calculate the rate constant, k, for this reaction. State the units for k. [3]

  b Bromine can **also** be formed by the oxidation of hydrogen bromide with oxygen.

  The following mechanism has been suggested for this multi-step reaction.

step 1	$HBr + O_2 \rightarrow HBrO_2$
step 2	$HBrO_2 + HBr \rightarrow 2HBrO$
step 3	$HBrO + HBr \rightarrow Br_2 + H_2O$
step 4	$HBrO + HBr \rightarrow Br_2 + H_2O$
	(a repeat of step 3)

  (i) Explain the term *rate-determining step*. [1]
  (ii) The rate equation for this reaction is:
  rate $= k[HBr][O_2]$.
  Explain which of the four steps above is the **rate-determining step** for this reaction. [2]

(iii) Determine the **overall** equation for this reaction. [1]

[Total: 14]

OCR, June 2003

4 Hydrogen chloride is used in the manufacture of many chemical compounds, including those used in metallurgy and food processing.

a There are two main industrial methods for preparing hydrogen chloride:
- by direction combination of chlorine and hydrogen gases,
- as a by-product of the chlorination of many organic hydrocarbons.

Write equations to show the formation of HCl from:
(i) chlorine and hydrogen [1]
(ii) chlorine and hexane, C_6H_{14}. [1]

b Hydrochloric acid is usually sold as a solution prepared by dissolving hydrogen chloride gas in water.

A science technician bought $15.0\,dm^3$ of $8.00\,mol\,dm^{-3}$ hydrochloric acid which had been made by dissolving hydrogen chloride gas in water.

1 mol of gas molecules occupies $24.0\,dm^3$ at room temperature and pressure, r.t.p.

(i) Calculate the volume of hydrogen chloride gas at r.t.p. that dissolved to produce this hydrochloric acid. [2]
(ii) Outline, with quantities, how the technician could make up $1.00\,dm^3$ of $0.0200\,mol\,dm^{-3}$ hydrochloric acid from the $8.00\,mol\,dm^{-3}$ stock solution of hydrochloric acid. [2]
(iii) Calculate the pH of $0.0200\,mol\,dm^{-3}$ HCl(aq). [2]

c Hydrochloric acid can be neutralised with aqueous ammonia to form ammonium chloride

$$NH_3(aq) + HCl(aq) \rightarrow NH_4Cl(aq)$$

The technician titrated the $0.0200\,mol\,dm^{-3}$ hydrochloric acid prepared in **b**(ii) with aqueous ammonia.

A $20.0\,cm^3$ sample of the $0.0200\,mol\,dm^{-3}$ HCl(aq) was placed in a conical flask and the $NH_3(aq)$ was added from a burette until the pH no longer changed. The pH curve for this tritration is shown below.

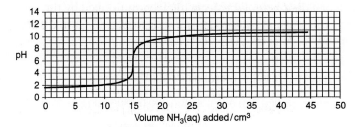

(i) How can you tell from this pH curve that aqueous ammonia is a weak base? [1]
(ii) Use the information above to calculate the concentration, in $mol\,dm^{-3}$, of the aqueous ammonia. [2]

(iii) The pH ranges in which the pH changes for three indicators are shown opposite.

Indicator	pH range
Alizarin yellow	10.1–12.0
Methyl yellow	2.9–4.0
Chlorophenol red	4.8–6.4

Explain which of the three indicators is most suitable for this titration. [2]

[Total: 13]

OCR, June 2003

5 *Syngas* is a mixture of carbon monoxide and hydrogen gases, used as a feedstock for the manufacture of methanol.

A dynamic equilibrium was set up between carbon monoxide, CO, hydrogen, H_2, and methanol, CH_3OH. The equilibrium system is shown below.

$$CO(g) + 2H_2(g) \rightleftharpoons CH_3OH(g)$$

The equilibrium concentrations of the three components of this equilibrium are shown below.

Component	CO(g)	H₂(g)	CH₃OH(g)
Equilibrium concentration /mol dm⁻³	3.1×10^{-3}	2.4×10^{-2}	2.6×10^{-5}

a State **two** features of a system that is in *dynamic equilibrium*. [2]

b (i) Write the expression for K_c for this equilibrium system. [2]
(ii) Calculate the numerical value of K_c for this equilibrium. [2]

c The pressure was increased whilst keeping the temperature constant. The system was left to reach equilibrium. The equilibrium position shifted to the right.
(i) Explain why the equilibrium moved to the right. [2]
(ii) What is the effect, if any, on K_c? [1]
(iii) State and explain the effect on the rates of the forward and reverse reactions
- when the pressure was first changed
- when the system reached equilibrium. [4]

d The temperature was increased whilst keeping the pressure constant. The system was left to reach equilibrium. The value of K_c for the equilibrium decreased.
(i) Explain what happens to the equilibrium position. [2]
(ii) Deduce the sign of the enthalpy change for the forward reaction shown in the equilibrium. Explain your reasoning. [1]
(iii) Explain how the partial pressure of $CH_3OH(g)$ would change as the system moved towards equilibrium. [1]

[Total: 17]

OCR, Jan 2003

14

Mathematics in chemistry

14.1 Numbers expressed as powers of ten

A shorthand way of writing very large or very small numbers is to arrange them so that they are written as powers of ten.

1×10^0 means 1 (Any number raised to the power 0 is 1)
1×10^1 means 10
1×10^2 means $10 \times 10 = 100$
1×10^3 means $10 \times 10 \times 10 = 1000$
1×10^4 means $10 \times 10 \times 10 \times 10 = 10\,000$

It works for numbers less than 1 too.

1×10^{-1} means $\dfrac{1}{10} = 0.1$

1×10^{-2} means $\dfrac{1}{100} = 0.01$

1×10^{-3} means $\dfrac{1}{1000} = 0.001$

To multiply these numbers, we *add* the powers, and to divide them we *subtract* the powers.

For example:

$10^2 \times 10^3 = 100 \times 1000 = 10^5$

$\dfrac{10^3}{10^4} = \dfrac{1000}{10000} = \dfrac{1}{10} = 10^{-1}$

Work out the following:

a $1 \times 10^2 \times 1 \times 10^{21}$ **b** $1 \times 10^5 \times 1 \times 10^{-3}$ **c** $\dfrac{10^0}{10^3}$

Answers   **a** 1×10^{23} **b** 1×10^2 **c** 1×10^{-3}

14.2 Standard form

Writing a number with one digit before the decimal place, multiplied by the correct powers of ten, is called using standard form, see *Essential AS Chemistry for OCR*, Mathematics appendix.

Examples

The Avogadro constant (602 300 000 000 000 000 000 000) is 6.023×10^{23}.

The mass of a hydrogen atom (0.000 000 000 000 000 000 000 000 001 6) is 1.6×10^{-27} kg.

The charge on an electron (0.000 000 000 000 000 000 16) is $-1.6. \times 10^{-19}$ C.

The speed of light (300 000 000) is 3×10^8 m s^{-1}.

To multiply numbers written in standard form you multiply the numbers and *add* the powers of ten.

For example $(2 \times 10^9) \times (4 \times 10^{11}) = 2 \times 4 \times 10^{20} = 8 \times 10^{20}$

To divide numbers written in standard form, divide the numbers and subtract the powers.

For example $\dfrac{2 \times 10^9}{4 \times 10^{11}} = \times 10^{-2} = 0.5 \times 10^{-2} = 5.0 \times 10^{-3}$

14.3 Using logarithms

A logarithm is the value of the power to which 10 must be raised to give the number.

The logarithm of a number to the base 10 is written $\log_{10}$ or just log. One use of logarithms is as a method for scaling down very large numbers. The pH scale is a scale based on logarithms and the Henderson equation, which is about buffers, also uses logarithms.

You don't need to understand how logarithms work to use them; an outline of the theory follows for interest.

It is very straightforward to find the logarithm for powers of 10, see Table 14.1.

Table 14.1 *Logarithms of powers of 10. Notice that the logs of numbers that are less than 1 have a negative sign*

Number	Written as a power of 10	Logarithm
0.01	10^{-2}	$\log_{10} 0.01 = -2$
0.1	10^{-1}	$\log_{10} 0.1 = -1$
1	10^0	$\log_{10} 1 = 0$
10	10^1	$\log_{10} 10 = 1$
100	10^2	$\log_{10} 100 = 2$
1000	10^3	$\log_{10} 1000 = 3$
10000	10^4	$\log_{10} 10\,000 = 4$

> **NOTE**
>
> We also use $\log_e$ of a number, where e is the number 2.71828 (to 5 decimal places). This is usually ln on your calculator. Don't confuse this with $\log_{10}$.

But, *any* number can be written as 10^x in which case $\log_{10} 10^x = x$ and we can find x using tables or calculators. For example, $\log_{10} 2$ is 0.3010 (which means that 2 is actually $10^{0.3010}$).

We can also work from the log to the number. This is called finding the antilog. The antilog is the number that gave us the log. In the example above, 2 is the antilog of 0.3010.

You can see from the table that if the log is a whole number, i.e. 1, 2, 3, etc., the antilog will simply be 10 to the power of the log. For example, if the log is -3.00, then the antilog is 10^{-3}. If the log is 9.00, then the antilog is 10^9. Any other antilog must be found from tables or calculators, see Section 14.1.5.

Work out the following without using tables or a calculator:

a $\log_{10} 0.001$,   **b** $\log_{10} 1\,000\,000$,   **c** the number whose log = 30,
d the number whose log = -4.

Answers   **a** -3 **b** 6   **c** 1×10^{30} **d** 1×10^{-4}

14.4 Multiplying and dividing using logarithms

There are two rules:

- To multiply two numbers together we add their logs and then find the antilog of the sum.

So: to calculate $Z \times Y$, we add $\log Z + \log Y$ and then find the antilog of the sum.

133

- To divide two numbers we subtract the logs and then find the antilog of the difference:

So: to calculate $\dfrac{Z}{Y}$, we subtract log Z – log Y and then find the antilog of the difference.

You can see why this works if we use powers of ten to demonstrate the principle.

Multiply $10^3 \times 10^4$ and you will get $10 \times 10 \times 10 \times 10 \times 10 \times 10 \times 10 = 10^7$

Using logs: $\log 10^3 = 3$ and $\log 10^4 = 4$

So, $\log 10^3 + \log 10^4 = 7$. The antilog of 7 is 10^7.

Divide 10^3 by 10^4 and you will get $\dfrac{10 \times 10 \times 10}{10 \times 10 \times 10 \times 10} = \dfrac{1}{10} = 10^{-1}$

Using logs: $\log 10^3 = 3$ and $\log 10^4 = 4$

So, $\log 10^3 - \log 10^4 = -1$. The antilog of -1 is 10^{-1}.

Write down the logs of the following numbers and use the logs to do the calculation:

a $10^5 \times 10^6$ **b** $10^7 \times 10^{-4}$ **c** $\dfrac{10^6}{10^3}$ **d** $\dfrac{10^2}{10^{-6}}$

Answers **a** $5 + 6$, 10^{11} **b** $7 + (-4)$, 10^3 **c** $6 - 3$, 10^3 **d** $2 - (-6)$, 10^8

Glossary

Carbonyl compound: An organic compound which has the functional group $C=O$.

d-block element: An element whose highest energy electrons are in d-orbitals.

Dipole–dipole forces: Intermolecular forces caused by the attraction between permanent dipoles ($\delta+$ and $\delta-$ areas on molecules).

Electronegativity: The ability of an atom to attract the electrons in covalent bonds towards itself.

Electrophile: A reagent in organic chemistry that attacks electron-rich areas of molecules.

Endothermic: Describes a reaction in which heat is taken in as the reactants change to products – the temperature thus drops.

Enthalpy: Energy measured under conditions of constant pressure and at a specified temperature (the conditions are normally 100 kPa and 298 K).

Exothermic: Describes a reaction in which heat is given out as the reactants change to products – the temperature thus rises.

Functional group: In organic chemistry a reactive group attached to a hydrocarbon chain.

Hydrogen bond: An intermolecular force that acts between a strongly electonegative atom (fluorine, oxygen or nitrogen) and a hydrogen atom that is covalently bonded to an electronegative atom).

Indicator: A solution whose colour changes, depending on the pH of the solution to which it is added.

Inductive effect: The electron-releasing effect of alkyl groups such as $-CH_3$ or $-C_2H_5$.

Locant: A number used in the systematic name of an organic compound that shows to which atom of the main hydrocarbon chain a substituent is attached.

Lone pair: A pair of electrons in the outer shell of an atom or ion that is not shared with another atom.

Nucleophile: A reagent in organic chemistry that attacks positively charged carbon atoms.

Period: A horizontal row of elements in the Periodic Table. There are trends in the properties of the elements as we cross a period.

Polarising: Distorting the electron cloud surrounding an atom or ion so that it acquires a dipole.

Protonated: A species is protonated if it has a H^+ ion bonded to it.

Structural formula: The formula of an organic compound written so that each carbon is shown separately along with the other atoms bonded to it.

Transition element: An element that forms at least one compound in which it has a part-full d-shell of electrons.

Unsaturated: This describes an organic compound with one or more carbon–carbon double (or triple) bonds.

Volatile: This describes a substance that easily forms a vapour.

The Periodic Table of the Elements

Key

relative atomic mass
atomic symbol
name
atomic number

Group 1	2											3	4	5	6	7	0
							1.0 **H** hydrogen 1										4.0 **He** helium 2
6.9 **Li** lithium 3	9.0 **Be** beryllium 4											10.8 **B** boron 5	12.0 **C** carbon 6	14.0 **N** nitrogen 7	16.0 **O** oxygen 8	19.0 **F** fluorine 9	20.2 **Ne** neon 10
23.0 **Na** sodium 11	24.3 **Mg** magnesium 12											27.0 **Al** aluminium 13	28.1 **Si** silicon 14	31.0 **P** phosphorus 15	32.1 **S** sulphur 16	35.5 **Cl** chlorine 17	39.9 **Ar** argon 18
39.1 **K** potassium 19	40.1 **Ca** calcium 20	45.0 **Sc** scandium 21	47.9 **Ti** titanium 22	50.9 **V** vanadium 23	52.0 **Cr** chromium 24	54.9 **Mn** manganese 25	55.8 **Fe** iron 26	58.9 **Co** cobalt 27	58.7 **Ni** nickel 28	63.5 **Cu** copper 29	65.4 **Zn** zinc 30	69.7 **Ga** gallium 31	72.6 **Ge** germanium 32	74.9 **As** arsenic 33	79.0 **Se** selenium 34	79.9 **Br** bromine 35	83.8 **Kr** krypton 36
85.5 **Rb** rubidium 37	87.6 **Sr** strontium 38	88.9 **Y** yttrium 39	91.2 **Zr** zirconium 40	92.9 **Nb** niobium 41	95.9 **Mo** molybdenum 42	– **Tc** technetium 43	101 **Ru** ruthenium 44	103 **Rh** rhodium 45	106 **Pd** palladium 46	108 **Ag** silver 47	112 **Cd** cadmium 48	115 **In** indium 49	119 **Sn** tin 50	122 **Sb** antimony 51	128 **Te** tellurium 52	127 **I** iodine 53	131 **Xe** xenon 54
133 **Cs** caesium 55	137 **Ba** barium 56	139 **La** lanthanum 57 *	178 **Hf** hafnium 72	181 **Ta** tantalum 73	184 **W** tungsten 74	186 **Re** rhenium 75	190 **Os** osmium 76	192 **Ir** iridium 77	195 **Pt** platinum 78	197 **Au** gold 79	201 **Hg** mercury 80	204 **Tl** thallium 81	207 **Pb** lead 82	209 **Bi** bismuth 83	– **Po** polonium 84	– **At** astatine 85	– **Rn** radon 86
– **Fr** francium 87	– **Ra** radium 88	– **Ac** actinium 89 *	– **Rf** rutherfordium 104	– **Db** dubnium 105	– **Sg** seaborgium 106	– **Bh** bohrium 107	– **Hs** hassium 108	– **Mt** meitnerium 109	– **Uun** ununnilium 110	– **Uuu** unununium 111	– **Uub** ununbium 112		– **Uuq** ununquadium 114		– **Uuh** ununhexium 116		– **Uuo** ununoctium 118

lanthanides *

140 **Ce** cerium 58	141 **Pr** praseodymium 59	144 **Nd** neodymium 60	– **Pm** promethium 61	150 **Sm** samarium 62	152 **Eu** europium 63	157 **Gd** gadolinium 64	159 **Tb** terbium 65	163 **Dy** dysprosium 66	165 **Ho** holmium 67	167 **Er** erbium 68	169 **Tm** thulium 69	173 **Yb** ytterbium 70	175 **Lu** lutetium 71

actinides * *

– **Th** thorium 90	– **Pa** protactinium 91	– **U** uranium 92	– **Np** neptunium 93	– **Pu** plutonium 94	– **Am** americium 95	– **Cm** curium 96	– **Bk** berkelium 97	– **Cf** californium 98	– **Es** einsteinium 99	– **Fm** fermium 100	– **Md** mendelevium 101	– **No** nobelium 102	– **Lr** lawrencium 103

Answers to quick questions

Chapter 1

1.1
1 CH
2 **a** $C_{10}H_8$ **b** C_5H_4
3 **a** 3 **b** 4
   c The ring structure of benzene means 1 less hydrogen molecule is needed.
4 This means that some of the bonding electrons are spread out over all six carbon atoms in the ring.

1.2
1 The polarity of a bond depends on the difference in electronegativity of the two atoms in the bond. This is zero for the C—C bonds and small (0.4) for the C—H bonds.
2 van der Waals' forces
3 **a** 1,2-dichlorobenzene **b** 1,4-dichlorobenzene
4 **a**

   b
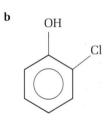
5 H^+

1.3
1 $C_6H_6(l) + 7\frac{1}{2}O_2(g) \rightarrow 6CO_2(g) + 3H_2O(l)$
2 Chlorobenzene
3 Chlorine is the most electronegative halogen and draws electrons away from the benzene ring most effectively.
4 Addition reactions would destroy the aromatic system.

1.4
1 (i) Electrophilic substitution (ii) electrophilic substitution
2
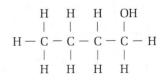
3 $C_2H_5^+$
4 1,2-dinitrobenzene and 1,4-dinitrobenzene

1.5
1 **a** $C_6H_5OH + NaOH \rightarrow C_6H_5ONa + H_2O$
   b $C_6H_5OH + Na \rightarrow C_6H_5ONa + \frac{1}{2}H_2$
2 The reaction with sodium hydroxide
3 Tertiary

---

4 **a**
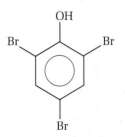

   b Hydrogen bromide, HBr
   c Liquid bromine with an iron catalyst
   d Liquid rather than bromine solution is required. Only one bromine atom substitutes rather than three. A catalyst is required.

1.6
1 2,4,6-trichlorophenol
2 Phenol has a polar —O—H group, which can form hydrogen bonds with water.
3 Phenol is volatile (it vaporises easily) and so can be smelled. It reacts with sodium hydroxide to form the involatile ionic compound sodium phenoxide. When acid is added, the phenol reforms.
4 **a** Carbon
   b It has a higher percentage of carbon in its molecule.

Chapter 2

2.1
1 **a** Pentan-2-one **b** propanal
2 **a** A ketone must have a C=O group with two carbon-containing groups attached to it.
   b The C=O group must be in the body of the chain. Only one position (the 2-position) is possible.
   c The C=O group must be at the end of the chain.
3 A hydrogen bond needs a molecule with a hydrogen atom covalently bonded to O, N or F. There is no such hydrogen in propanone.
4 The O—H group in a water molecule can hydrogen bond with the O of the C=O in propanone.

2.2
1 $:Cl^-$
2 **a** No
   b $:H^-$ will not attack the electron-rich C=C.
   c
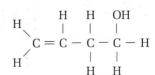
3
H—C—C—C—C—H structure

137

Chapter 3

3.1
1   3-bromobutanoic acid
2

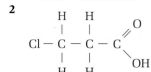

3   The carboxylic acid group can only be on the end of the chain.
4   They will react with metal hydroxides, metal oxides and metal
    carbonates.

3.2
1   $HCOOH + Li \rightarrow HCOOLi + \frac{1}{2}H_2$
2   $2CH_3CH_2COOH + MgO \rightarrow (CH_3CH_2COO)_2Mg + H_2O$
3   Ethanoic acid and methanol
4   Methanoic acid and ethanol
5   They have the same molecular formula but different structure.

Chapter 4

4.1
1   Secondary
2   Ethylpropylamine
3

$$CH_3$$
$$|$$
$$H_3C - N - CH_3$$

4   A gas
5   It is an isomer of ethylamine, which is a gas.

4.2
1   a $(CH_3)_2NH + HCl \rightarrow (CH_3)_2NH_2^+ + Cl^-$
    b   Dimethylammonium chloride, dimethylamine hydrochloride
2   a   Oily drops disappear.
    b   Oily drops reappear.
3   Stronger. It has two alkyl groups releasing electrons on to the
    nitrogen atom.

4.3
1   They are strongly coloured and bond well to cloth.
2   An electrophile
3   Electrons in π-bonds are spread over three or more atoms.
4   180°

4.4
1   a   Amine, $-NH_2$ and carboxylic acid $-COOH$ (b) Amine is
        basic, carboxylic acid is acidic.
2   2-aminopropanoic acid
3   Two
4   a $-NH_2$,   b   the C of $-COOH$

Chapter 5

5.1
1   d $RCH=CHR$
2   a   *trans*-pent-2-ene   b   *cis*-pent-2-ene
3   b
4

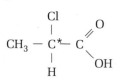

5.2
1   a   2-hydroxybutanoic acid
    b   Yes, it has a carbon atom with four different groups
        attached.
2   a   2-hydroxy-2-methylpropanoic acid
    b   No, it has no carbon with four different groups attached
3   Two of the groups attached to it are the same (methyl groups)

Chapter 6

6.1
1   All except **c**
2   a

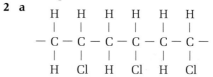

    b   chloroethene
    c   poly(chloroethene)
3   Syndiotactic. The benzene group is alternately above and below
    the plane of the polymer.

6.2
1   a   From the number of carbon atoms in each monomer.
    b   1,10-diaminodecane, $H_2H(CH_2)_{10}NH_2$
2   They also have many $-CONH-$ (amide) linkages.
3   Any other diol, e.g. propane-1,3-diol

6.3
1   a CF_2 b $CF_2=CF_2$
2   a CH_2CHCl b $CH_2=CHCl$
3   a   Addition polymerisation. The monomer has a $C=C$.
4   a   Amide   b   ester

Chapter 7

7.2
1   a   (i)   ~1725 cm^{-1}
        (ii) ~3000 cm^{-1}
    b   (i) $C=O$ (ii) $C=O$ (iii) $C=O$ (iv) $O-H$ (v) $O-H$
2   Virtually all organic compounds have $C-H$ bonds.
3   In pure ethanol the $O-H$ groups will hydrogen bond to one or
    more other ethanol molecules. This is much less likely in a
    solvent, which cannot form hydrogen bonds with ethanol.
4   The mass of the atoms involved in these two bonds are very
    similar (and the bond strengths are also similar).

7.3
1   a   (i)   120 (ii) 105
    b   120
    c   That it has a single (positive) charge.
2   a   Electrons are knocked out by bombardment in an electron
        beam.
    b   Positive
3   a   (i)   The molecular ion may break up (fragment).
        (ii) These are caused by isotopes, such as carbon-13.
    b   The abundance of heavier isotopes is usually low.

7.4
1   a   Flip to line up as in **c**.
    b   Flip to line up as in **c**.
    c   no movement
2   a CH_3CH_2OH
    b   3
    c   There are hydrogen atoms in three different environments.
    d   3:2:1
    e   There are 3 hydrogens in the CH_3 group, 2 in the CH_2 group
        and 1 in the OH group.

3 **a** One
  b All the hydrogen atoms are in an identical environment.

7.5

1 They have the same molecular formula but different structures.
2

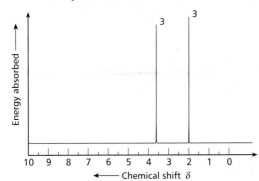

propan-1-ol   propan-2-ol

  a (i) 4   (ii) 3
  b (i) Propan-1-ol, A = 3, B = 2, C = 2, D = 1
    (ii) propan-2-ol, A = 6, B = 1, C = 1.
  c (i) Propan-1-ol, A = 0.7–1.6, B = 1.2–1.4, C = 3.3–4.3,
      D = 3.5–5.5
    (ii) propan-2-ol, A = 0.7–1.6, B = 3.3–4.3, C = 3.5–5.5.

7.6

1 **a** A is ethanol; B is methoxymethane.
  b A. 4.5 is R—O—H, 3.7 is R—CH_2—O—, 1.2 is R—CH_3.
    B. 3.3 is —O—CH_3.
  c A. 4.5, 1; 3.7, 2; 1.2, 3.
    B. 3.3 not possible to tell.
2

[Graph: Energy absorbed (y-axis) vs Chemical shift δ (x-axis), with peaks of height 3 at δ ≈ 3.7 and δ ≈ 1.2; x-axis labelled 10 9 8 7 6 5 4 3 2 1 0]

Note: There is no spin–spin splitting because the two CH_3 groups are not on adjacent carbons.

7.7

1 **a** The 3 hydrogens at $\delta = 2.1$ represent the CH_3 group.
    The hydrogen at $\delta = 11.7$ represents the —O—H group.
  b The peak at $\delta = 11.7$ would disappear.
2 **a** (i) $CH_3CH_2CH_2OH$ (ii) HCHOCHO
  b (i) $CH_3CH_2CH_2OH$
  c because it has an —OH group.
3 **a** Yes, an —OH group is present.
  b Yes, an —OH group is present.
  c No, neither —OH nor —NH groups are present.
  d No, neither —OH nor —NH groups are present.
  e Yes, an —OH group is present.

Chapter 8

8.1

1 **a**

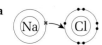

  b 1st ionisation energy of sodium, 1st electron affinity for chlorine.
  c Atomisation enthalpies of sodium and of chlorine, enthalpy of formation of sodium chloride, lattice enthalpy of sodium chloride

2 **a** The negatively charged electron has to be pulled away from the positive charge of the nucleus.
  b The electron that is gained is attracted by the positively charged nucleus of the chlorine.
3 **a** (i) $Mg(g) \rightarrow Mg^+(g) + e^-$
    (ii) $Mg^+(g) + e^- \rightarrow Mg^{2+}(g) + e^-$
  b The sum of the two enthalpy changes above.

8.2

1 **a**

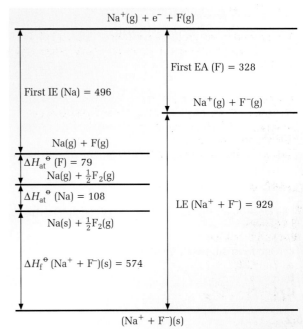

  b -929 kJ mol^{-1}

8.3

1 The Li^+ ion is smaller than Na^+ so when the lattice forms the F^- ion can approach the metal ion more closely in LiF than NaF.
2 The Cl^- ion is larger than F^- so when the lattice forms the Cl^- cannot approach the Li^+ as closely as LiCl as in LiF.
3 In MgO (Mg^{2+} O^{2-}) both ions have double charge compared with the ions in NaF (although they are approximately the same size).
4 The metal ions are of comparable size because they are from Period 3 of the Periodic Table, and the non-metal ions are of comparable size, because they are both in Period 2.

Chapter 9

9.1

1 Conduct heat well, malleable (can be beaten into shape), ductile (can be drawn into a wire), sonorous (will ring when hit with a hammer), form basic oxides.
2 Conduct heat poorly (are insulators), form acidic oxides, tend to be brittle.
3 Mg^{2+} O^{2-}
4 **a**

  b Conducts electricity when molten.
  c High melting and boiling temperatures.
5 **a** Never conducts electricity
  b Low melting and boiling temperatures

9.2

1 **a** The proportion of chlorine atoms in the compound increases.
   b SCl_6
2 Chlorine cannot react with itself. Argon is a noble gas.
3 $(1 \times +II) + 2 \times (-II + I) = 0$

9.3

1 **a** $Na_2O(s) + H_2O(l) \rightarrow 2NaOH(aq)$
   b (i) +I before and after, (ii) neither
2 They become less soluble.
3 Show that they conduct electricity.
4 **a** OH^-
   b Greater than 7
5 **a** H^+
   b Less than 7

Chapter 10

10.1

1 **a** (i) $1s^2\,2s^2\,2p^6\,3s^2\,3p^6\,3d^5$
        (ii) $1s^2\,2s^2\,2p^6\,3s^2\,3p^6\,3d^4$
   b It has a half-full d-shell.
2 The electron arrangement of argon; $1s^2\,2s^2\,2p^6\,3s^2\,3p^6$
3 **a** $Sc \rightarrow Sc^{3+} + 3e^-$
   b 3
   c One from 3d, two from 4s.
4 **a** $Zn \rightarrow Zn^{2+} + 2e^-$
   b 2
   c 4s

10.2

1 **a** A homogeneous catalyst is in the same phase as the reactants and products. A heterogeneous catalyst is in a different phase from the reactants and products.
   b (i) Heterogeneous,
       (ii) heterogeneous,
       (iii) homogeneous
2 H^+.
3 One in which both the shared electrons come originally from the same atom.

10.3

1 $90°$
2 $109.5°$
3 A catalyst speeds up a reaction but is not used up.
4 It has an extra full shell of electrons compared with oxygen and nitrogen.
5     $+II$ $-II+I$ $+II$ $-II+I$
   a $Cu^{2+}(aq) + 2OH^-(aq) \rightarrow Cu(OH)_2(s)$
   b No. None of the oxidation numbers changes.

10.4

1 **a** (i) $1s^2\,2s^2\,2p^6\,3s^2\,3p^6\,3d^6$
        (ii) $1s^2\,2s^2\,2p^6\,3s^2\,3p^6\,3d^5$
   b Fe^{3+}. It has a half-full shell of d-electrons.
2 It has a negative charge.
3 **a** $+III$ $-II +I$ $+III$ $-II+I$
       $Fe^{3+}(aq) + 3OH^-(aq) \rightarrow Fe(OH)_3(s)$
   b No. No oxidation numbers change.
4 More than enough to react.

10.5

1 It will look green – it absorbs blue and red and allows green to pass through.
2 Sc^{3+} $1s^2\,2s^2\,2p^6\,3s^2\,3p^6$; Zn^{2+} $1s^2\,2s^2\,2p^6\,3s^2\,3p^6\,3d^{10}$. Neither has part-full d-shell.

3 Red and blue pass through, green is absorbed.

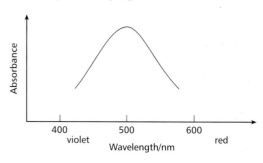

4 **a** 5:5   **b** 5:5   **c** $[NiEDTA]^{2+}$

10.6

1 **a** (i) Fe +III, O −II, C 0   (ii) Fe 0, O −II, C +II
   b C   **c** Fe
2 $Zn(s) + 2V^{3+}(aq) \rightarrow Zn^{2+}(aq) + 2V^{2+}(aq)$
3 0.069 g

Chapter 11

11.1

1 A reactant – its concentration decreases with time.
2 $0.09\ mol\ dm^{-3}\ min^{-1}$ to one significant figure
3 The rate of reaction after 5 minutes.
4 The gradient at time = 0 minutes will be steeper than that at time = 5 minutes. The gradient at time = 10 minutes will be less steep.
5 The rate of reaction is fast at the beginning when there is plenty of reactant and slower as the reactant is used up.

11.2

1 Rate = $k[A][B][C]^2$
2 **a** (i) 1   (ii) 1   (iii) 2
   b (i) double   (ii) double   (iii) quadruple
   c (i) 1   (ii) 5   (iii) 6   (iv) 3   (v) 3
   d $dm^9\ mol^{-3}\ s^{-1}$
3 **a** The rate constant
   b (i) 2   (ii) 0   (iii) 0   (iv) 1
   c 3
   d $dm^6\ mol^{-2}\ s^{-1}$
   e It is a catalyst.
4 d

11.3

1 **a** Reactant. Its concentration decreases with time.
   b Zero order. A zero order reaction would have a straight line graph.
   c Measure three (or more) successive half-lives. They will be the same if the reaction is first order.
2 **a** Zero order. A zero order reaction would have a straight line graph.
   b (i) They will be the same.   (ii) They will increase.
   c Second order – the half lives increase.

11.4

1 **a** (i) 1   (ii) 2
   b 3
   c 27
   d Rate = $k[A][B]^2$
   e No, species other than A and B may also be involved.

11.5

1 c
2 **a** B
   b E
   c step (ii)

Chapter 12

12.1

1 a $K_c = \dfrac{[C]_{eqm}}{[A]_{eqm}[B]_{eqm}}$

  b $K_c = \dfrac{[C]_{eqm}}{[A]^2_{eqm}[B]_{eqm}}$

  c $K_c = \dfrac{[C]^2_{eqm}}{[A]^2_{eqm}[B]^2_{eqm}}$

2 a dm^3mol^{-1}
  b dm^6mol^{-2}
  c dm^6mol^{-2}

3 a 3.5   **b** It cancels out   **c** to the left

12.2

1 a $x = 1.01$ mol   **b** 1.03 mol   **c** 2.07 mol

12.3

1 $pI_2(g)_{eqm} = 10.99$ kPa, $pHI(g)_{eqm} = 78.02$ kPa

2 a $\dfrac{pSO_3(g)_{eqm}^2}{pSO_2(g)_{eqm}^2 \, pO_2(g)_{eqm}}$

  b $\dfrac{pNO_2(g)_{eqm}^2}{pNO_2O_4(g)_{eqm}}$

  c $\dfrac{pH_2O(g)_{eqm}\, pCO(g)_{eqm}}{pH_2(g)_{eqm}\, pCO_2(g)_{eqm}}$

3 a kPa^{-2} **b** kPa   **c** no units

12.4

1 a (i) move to the right   (ii) move to the left
  b (i) move to the left   (ii) move to the right
  c (i) no change   (ii) move to the right

Chapter 13

13.1

1 a Br^- **b** H_2O
2 HNO_3 is an acid and NO_3^- its conjugate base. OH^- is a base and H_2O its conjugate acid.
3 An alkali is a water-soluble base.
4 10^{-10} mol dm^{-3}
5 a H_2O **b** NH_4^+ **c** H_3O^+ **d** HCl.

13.2

1 2
2 1×10^{-6} mol dm^{-3}
3 1×10^{-5} mol dm^{-3}
4 1.69
5 13.3

13.3

1 Chloroethanoic acid – it has the largest value of K_a.
2 0
3 a 1.94
  b 3.10
4 They are the same.

13.5

1 It changes colour gradually over many pH units.
2 **b** and **d**

13.6

1 a 5.1
  b 4.2

Index